SpringerBriefs in Applied Sciences and Technology

T0192300

More information about this series at http://www.springer.com/series/8884

Jenny M. Jones · Amanda R. Lea-Langton
Lin Ma · Mohamed Pourkashanian
Alan Williams

Pollutants Generated by the Combustion of Solid Biomass Fuels

 Springer

Jenny M. Jones
Amanda R. Lea-Langton
Energy Research Institute
University of Leeds
Leeds, UK

Lin Ma
Mohamed Pourkashanian
Alan Williams
Energy Technology and Innovation
    Initiative
University of Leeds
Leeds, UK

ISSN 2191-530X
ISBN 978-1-4471-6436-4
DOI 10.1007/978-1-4471-6437-1

ISSN 2191-5318   (electronic)
ISBN 978-1-4471-6437-1   (eBook)

Library of Congress Control Number: 2014953258

Springer London Heidelberg New York Dordrecht

Printed on acid-free paper

Springer is part of Springer Science+Business Media (www.springer.com)

Foreword

Biomass presents a vast renewable resource that can provide food, energy and chemicals for the worlds' population. The demand for each of these depends on the future development of other technologies and, most importantly, other sources of energy such as nuclear or solar. Progress is also dependent on scientific and technological developments. Firstly, in combustion technology to reduce environmental damage locally, nationally and globally. Improved combustion units are required with higher efficiencies and requiring less maintenance. Biomass will be used for domestic heating purposes on a large scale for some time to come. In order to overcome these emissions problems there are two possible lines of attack. The first lies in the possibility of fuel modification where fuels improved by agronomy or by using genetically modified crops may offer significant advantages. The second involves intellegent fuel blending which may also be advantageous.

There are key issues relating to sustainability in order to provide a constant source of renewable biomass. There are numerous important factors, including the complete utilisation of ash, so that the plant nutrients are sustainable. This is dependent on economic factors and life cycle issues.

It seems likely that the really important developments lie in the large-scale use of biomass electricity production, for heating and chemical production. These processes can also be coupled with carbon capture with sequestration and carbon dioxide utilisation. Such applications present the opportunity for the removal of carbon dioxide from the atmosphere although the magnitude of undertaking this on a substantial scale is enormous. These developments are dependent on engineering solutions such as innovative boiler and furnace designs and intelligent control systems. This will be assisted by computer modelling including virtual biomass plant (CFD) design solutions.

Finally, it should be noted that we have not attempted to describe the history of the development of the understanding of the formation of pollutants from biomass. This is so closely bound up with our understanding of the combustion of gases, hydrocarbon liquids and especially coal that many developments result as a synergy of research activities. Here we have given mainly recent references that also include the preceding literature.

Preface

Recent projections suggest that the world population will be higher than previously estimated and might reach 11 billion by the end of the century (Science, 18 September 2014). The fastest increase is in Sub-Saharan Africa. This projection is higher than that used by the Intergovernmental Panel on Climate Change which assumes a population peak in 2015. Consequently, the population increase will have a greater impact on the amount of energy required for heating, cooking, electricity production, transport, agriculture and the manufacturing industries.

Biomass makes a significant contribution to world energy consumption at the present time, although much of this is for the traditional use of biomass mainly in developing countries. However, it is expected that there will be significant growth in the use of bioenergy using more advanced technology for electricity generation and the provision of heat. Indeed, IEA predicts that the use of bioheat will increase by 60 % by 2035. This will be driven by concerns over climate change and renewable energy policy initiatives by governments. Security of supply issues will play a role because of the wide geographical distribution of biomass. Advances in technology will aid transportation, fuel pre-processing and combustor design.

Issues about resource availability and sustainability are very important. Competition for land potentially leads to food poverty, hence the food versus fuel debate will become extremely important. However, bioenergy presents a number of opportunities for the utilisation of agricultural wastes as solid, liquid and gaseous fuels. The integration of agriculture and bioenergy is an important future requirement.

The combustion of solid biomass will play a major role in these developments but it results in the formation of pollutants which have an adverse effect on the health of the community and on the climate. At the present time there is sufficient concern about these aspects so as to promote more stringent legislation. If the amount of bioheat increases by 60 % over the next two decades, then greater pollution control measures will need to be applied.

In addition, there is the significance of emissions of carbon dioxide resulting from the combustion of biomass. Biomass is potentially almost carbon neutral depending on the agricultural methods. If carbon capture and storage can be applied, then this could reduce the concentration of carbon dioxide in the atmosphere, thus mitigating climate change which would be beneficial to the world as a whole.

Acknowledgments

We wish to thank a number of our colleagues for their assistance—Prof. Keith Bartle, Prof. Marie Teresa Baeza-Romero, Prof. Dominick Spracklen, Prof. Christopher Williams, Dr. Leilani Darvell, Dr. Emma Fitzpatrick, Mr. Ed Butt, Mr. Doug Hainsworth, Mr. Eddy Mitchell and Mr. Farooq Atika. We would also like to thank Advanced Flight Training Ltd, Sherburn in Elmet, UK for aerial photography. Some of the work described has been supported by the UK Supergen Bioenergy programme.

Contents

Acronyms

ASME	American Society of Mechanical Engineers
BS	British Standard
BC	Black Carbon
CEN	Comité Européen de Normalisation
CFD	Computational Fluid Dynamics
daf	Dry ash free
EC	Elemental Carbon
FTIR	Fourier Transform Infra-Red
HHV	Higher Heating Value
ISO	International Organisation for Standardization
LES	Large Eddy Simulation
OC	Organic carbon
PAH	Polycyclic Aromatic Hydrocarbon
pf	Pulverised Fuel
PKE	Palm Kernel Extruder
PM_1	Particulate matter less than 1 μm in diameter
PM_{10}	Particulate matter less than 10 μm in diameter
$PM_{2.5}$	Particulate matter less than 2.5 μm in diameter
PY-GC-MS	Pyrolysis-Gas Chromatograph-Mass Spectrometer
RANS	Reynolds-Averaged Navier Stokes
RCG	Reed Canary Grass
SOC	Soil Organic Carbon
SRC	Short Rotation Willow Coppice
SW	Switchgrass
TGA	Thermal Gravimetric Analyser
TLV	Threshold Limit Value-a measure of toxicity
toe	Tons Oil Equivalent
UBH	Unburned hydrocarbons
VM	Volatile Matter

Nomenclature

A	Pre-exponential factor for devolatilisation
A_p	Surface area of the particle
A_i	Pre-exponential factor
B	Biot Number
C_g	Reacting gas species concentration in the bulk gas
C_1	Experimentally obtained diffusion coefficient
C_{pM}	Specific heat of moisture
D_o	Bulk diffusion rate coefficient
d_p	Particle diameter
D_s	Energy transfer due to species diffusions
E	Activation energy for devolatilisation
E_i	Activation energy for the intrinsic reactivity
F^i	Source term for an external force
F	X_m/M
g^i	Gravitational acceleration
ΔH_M	Enthalpy of vaporisation of moisture
K	Reaction rate constant/turbulent kinetic energy
k_{eff}	Effective rate constant/effective heat conductivity of fluid flow
Kgas	Thermal conductivity, gas
Ksolid	Thermal conductivity, solid
M	Initial moisture content
m_p	Particle mass
N	Dimensionless particle reaction order
Nu	Nusselt Number
P	Pressure
p_{ox}	Partial pressure of the oxidant species in the surrounding gas
R	Process rate/Gas Constant
R_c	R_c is the chemical kinetic reaction rate constant
S	Source term
T_p	Particle temperature
T_∞	Bulk gas temperature

T_{evap}	Evaporation temperature
t	Time
V	Volatile matter, non-dimensionalised
U	Fractional burnout
u^i, etc.	Mean fluid velocity components
X_i	Mol fraction of species i
X_M	Current moisture content

Greek Letters

ε	Dissipation rate
η	Effectiveness factor
ρ	Density
ρ_i	Intrinsic reactivity
μ	Molecular viscosity of fluid flow
τ^{ij}	Reynolds stress tensor
Φ	Related to viscous dissipation

Subscripts

e	Electrical
th	Thermal

Chapter 1
Introduction to Biomass Combustion

Abstract The role of biomass in the world energy scene is first considered and compared with other energy sources. Estimates of the resources available are provided and discussed. The IEA world estimates of the growth in the use of biomass are also presented. Issues resulting to Greenhouse Gas emissions and sustainability are considered.

Keywords Biomass availability · Resources · Greenhouse gases

1.1 The Role of Biomass Combustion in World Energy

Solid biomass consists of material of recent organic origin which has not yet undergone the process of metamorphosis that leads to peat and coal. It consists of three major components, cellulose, hemicellulose and lignin together with small amounts of solvent extractable organic materials as well as traces of inorganic compounds. These inorganic materials end up as ash and this term is commonly applied to this material before it is burned. Peat is not defined as a biomass. Charcoal is a very important derived biofuel (Antal and Gronli 2003). The boundary between biomass and waste is not well defined since the definition varies from country to country and can depend on local legislation. Here we include forestry and agricultural waste but not domestic, sewage or food industry waste although there may be many similarities in their properties.

Biomass was the major fuel in the world until the late 1,800 hundreds when coal then became dominant. The combustion of solid biofuels as a primary energy source has again grown significantly in the last decade principally because it can be used to replace the fossil fuels, coal, oil and natural gas. This is because it is a renewable energy resource and it offers neutrality or near-neutrality with respect to carbon dioxide emissions. It is derived from plant matter which is produced by photosynthesis from carbon dioxide in the atmosphere.

© The Author(s) 2014
J.M. Jones et al., *Pollutants Generated by the Combustion of Solid Biomass Fuels*,
SpringerBriefs in Applied Sciences and Technology,
DOI 10.1007/978-1-4471-6437-1_1

However, similar to fossil fuels there are some adverse consequences in its use as a fuel such as the formation of smoke, oxides of nitrogen and other environmental pollutants. Biofuels may arise from energy crops grown specifically for this purpose or are by-products from industrial process such as paper-making, or as agricultural wastes such as straw. This use as a fuel is in competition with biomass other uses such as the production of commodities such as food—the food versus fuel issue, or paper, fibreboard, furniture or bio-motor and aviation fuels, and as a feedstock for the chemical industry. A number of major reviews that have considered their utilisation and the environmental aspects of different processing technologies for example, Stevens and Brown (2011), Bridgwater et al. (2009), Spliethoff (2010).

The combustion of solid biomass thus covers a wide range of materials, woods, straw, agricultural residues processing wastes from paper manufacturing, and also algae and seaweeds. The broad use of the term biomass also includes related materials such as, manures, meat by-products and black liquor from paper making industries. Peat is not included but is similar from a combustion point of view and widely used as a fuel. Historically the major fuel has been wood; two centuries ago biomass provided almost 100 % of the energy used in the world and about 60 % a century ago. It is a major source of energy now in Africa, Asia and China. For example, biomass currently provides 78 % of the total energy in Nigeria.

At the present time biomass provides about 10 % of the world energy. Table 1.1 shows the current profile of energy use on a global basis in 2012. There are reasonably accurate statistics of the quantities of fossil produced and used (BP 2014) and also of renewable energy usage where the amounts used are reported. The main difficulty arises with the quantification of the large quantities of biomass used in the developing countries where records are not kept.

Of this the main application is in heating. Roughly two thirds is used in developing countries and the rest is in industrialised countries. The usage for steam raising and large electrical generation units is relatively small at about 15 %. The amount of liquid biofuels for transport is included for comparison and is small compared to the others (4 %). The approximate amounts used are shown in Table 1.2.

A number of analyses have been made of the future development of biomass but these predictions are complicated by political incentives such as carbon taxes and subsidies. The IEA have made the prediction of energy usage in the next 25 years using a scenario approach in their World Energy Outlook (2013) publication. The

Table 1.1 Global energy consumption in 2013

Fossil fuels oil, coal, natural gas	10.98 Gtoe
Nuclear	0.58 Gtoe
Hydro-electricity	0.86 Gtoe
Wind, solar, biomass	0.28 Gtoe
Biomass-not recorded, mainly in developing countries	1–1.5 Gtoe
Total-commercial Total-all uses	12.73 Gtoe 14 Gtoe approximately

Source For commercial recorded data: BP (2014)

Table 1.2 Main applications of biomass

Domestic cooking and heating. Heating of commercial premises-schools, large buildings	1 Gtoe
Industrial steam raising-waste biomass, medium size power generation/CHP/district heating units	0.19 Gtoe
Large electrical power plants	0.17 Gtoe
Transport	0.08 Gtoe

Source IEA Bioenergy (2013)

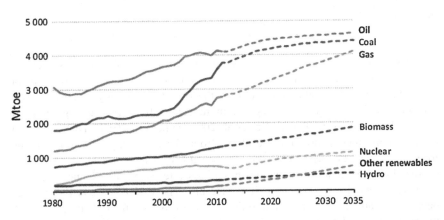

Fig. 1.1 IEA estimates of the world energy use until 2035 (World Energy Outlook 2013, with permission)

'New Policies Scenario' which is the middle of three options is shown in Fig. 1.1. The increase in future energy demand shows some levelling off in fossil fuel usage and an increase in the use of renewable energies especially biomass.

Predictions on the future usage of biomass have been made as far as to 2,100. It has been estimated that the contribution from bioenergy (liquid and solid fuels) could be raised to 200 EJ/year (4.8 Gtoe) by 2050 and possibly five times greater (IEA Bioenergy 2008).

Early biomass usage was based on open fires which had low efficiencies, typically 5–10 %, but this has significantly improved over the last hundred years. Local supplies were used which needed little in the way of processing, other than seasoning. This use developed widely especially in countries with considerable reserves of biomass, forested areas or bio-waste from sugar or paper industries, as shown in Table 1.3. This technology is still widely used in Developing Countries today despite major efforts to make other technologies available for cooking such as solar cookers. More recently an international trade has developed which is changing the world picture because of the need of fuels that are readily transported and stored without degradation or significant fire risk. A further factor in the present day application of biomass is that combustion requires much more consistent fuel quality and lower emission levels and so fuel pre-treatment may be necessary. This may involve pelleting or torrefaction which are described in Chap. 2.

Table 1.3   Use of solid biomass in the industrial sector in 2009 (IEA Bioenergy 2013)

Country	Primary solid biomass (Mtoe)	% Share of global use
Brazil	31	18
India	29	16
United States	25	16
Nigeria	9	5
Canada	7	4
Thailand	7	4
Indonesia	7	4

Today there is increased emphasis on the use of renewable energy sources in many industrialised countries and one of the options is biomass combustion for heat and electricity production. Bioheat, the oldest technology (Webster 1919; Tilman et al. 1981), is now moving to a more technically advanced stage in many industrialised countries (Bridgwater et al. 2009; Spliethoff 2010). This is a result of the environmental pressures resulting from climate change concerns arising from the release of GHG, and secondly because of security of energy supplies. Of the alternative renewable supplies biomass has fewer negatives since it is geographically widely available and it is storable. The major issue is the balance between the advantages presented in the fact that it is CO_2 neutral compared with the associated pollution, and the impact that it may have on food production and price. Some of the combustion applications result in a significant atmospheric pollution -indeed it is estimated that 1.5 million people die per annum as a result of this type of pollution (WHO 2014). Solid biomass fuels are in competition with the use as liquid biofuels or chemical feedstock, but the two could be integrated with the bio-waste from the processing application being used as solid fuel. The first generation of liquid biofuels were produced with minimal processing and low efficiency but the second generation has greater flexibility in the conversion into a variety of products. A number of factors have to be taken into account and the estimated GHG emissions associated with the bioenergy feedstocks or delivered energy can be estimated using life cycle analyses, and these are very important as well as economic factors. Thus in the UK the GHG emissions using various feedstocks and emissions savings compared with coal and gas were calculated for the UK by the UK Environment Agency (2009). The figures are listed in Table 1.4.

This type of information can be obtained for both supply and utilisation using Life Cycle Assessment (LCA) by means of a 'Biomass Emissions Calculator'. Such a Calculator was developed in the UK by DECC (The Department of Energy & Climate Change). These types of calculations are subject to a considerable amount of debate because of the assumptions made especially when it involves importing large quantities of biomass from North America and elsewhere for electricity generation. A more sophisticated calculator has been developed which looks at the changes in the amount of carbon stored in forests in North America when assessing the benefits and impacts of various bioenergy scenarios (Stephenson and MacKay

Table 1.4 GHG savings from various types of biomass feedstocks

	Feedstock	Emissions (kgCO$_2$e/MWh)
Chips	UK forestry residue	10
	Imported forestry residue	22
	Waste wood	7
	Short rotation coppice (SRC)	17
	Miscanthus	18
Pellets	UK forestry residues	38
	Imported forestry residue	50
	Waste wood	51
	Imported waste wood	66
	SRC	100
	Miscanthus	65
Other	Olive cake	9
	Palm Kernel expeller (PKE)	82
	Straw	73

2014). It gives information about which biomass resources are likely to have higher or lower carbon intensities, and so provides insight into a complex topic. In future this type of calculation will have to be made to make comparison with liquid biofuel production and other competing uses including the wood utilising industries. Like all models they are sensitive to the input data and the model assumptions.

1.2 Resources—Supply of Biomass

The current resource base for biomass fuels is about 1,000 Gtoe and is renewable and effectively infinite but only if the rate of growth is greater than the rate of usage. That is, if it is used in a sustainable way. The total amount of fuel available at some future time t, R_T, is given by:

$$R_T = [\text{The current resource base}] + \left[\text{The rate of growth}\right] - [\text{The rate of utilisation}]$$

But the rate of utilisation is the aggregate of the use of biomass for food, manufacturing industry (paper and furniture) and other process as well as a fuel. This is largely determined by the world population and whether this reaches an equilibrium level which is sustainable.

The usage of biomass is dependent the technology available, the environmental impact, the energy costs which are very dependent on subsidies paid by governments and ultimately the availability of the biomass at what is considered to be an economic price. Current and future estimates of biomass utilisation are subject to uncertainty and global values can vary by a factor of 5 but well-resourced information is available for major countries such as the US, Europe, Russia, China,

India and Brazil. The upper estimate of biomass annual availability (Smeets et al. 2007; Faaji and Londo 2010; Slade et al. 2011) is about 2,500 EJ (60 Gtoe) [1 EJ = 24 Mtoe] and is almost six times the world current energy requirement. Resources vary from country to country and are dependent on geographic location, the climate, the population density and the degree of industrialisation of the country.

It is seen from Table 1.1 that the present amount of energy derived from biomass is about 8–15 % of the total annual energy used. Essentially biomass used as a fuel can be divided into the following two major groups:

1. Biomass, grown directly for use as an industrial primary fuel,
2. Traditional bioenergy (e.g. biomass for cookstoves).

The amount of the former has been estimated as 6 EJ (144 Mtoe) and listed under "Renewables" in Table 1.1, and the latter as 39 EJ (936 Mtoe). This gives a total of 45 EJ (1,080 Mtoe) as indicated in Table 1.1, although some estimates make it slightly greater. Since the net global production of biomass is estimated as 2,280 EJ/year, so the percentage of the biomass used for fuel is 2 %. Food production on the other hand is equivalent to about 273 EJ/annum, that is approximately 12 %.

References

Antal MJ, Gronli MG (2003) The art, science, and technology of charcoal production. Ind Eng Chem Res 42:1619–1640

Bolling AK, Pagels J, Yttri KE, Barregard L, Sallsten G, Schwarze PE, Boman C (2009) Health effects of residential wood smoke particles: the importance of combustion conditions and physicochemical particle properties. Particle Fibre Toxicol 6:29

BP (2014) BP statistical review of world energy, June 2014 from www.bp.com

Bridgeman TG, Jones JM, Williams A (2010) Overview of solid fuels, characteristics and origin. In: Lackner M, Winter F, Agarwal AK (eds) Handbook of combustion. Solid fuels, vol 4. Wiley-VCH, pp 1–30

Bridgwater AV, Hofbauer H, van Loo S (eds) (2009) Thermal biomass conversion. CPL Press, Newbury

Environment Agency (2009) Minimising greenhouse gas emissions from biomass energy generation. Environment Agency, London

Faaji A, Londo M (eds) (2010) A road map for biofuels in Europe. Biomass Bioenergy 34:157–250

IEA Bioenergy (2008) The availability of biomass resources for energy. IEA Bioenegy:ExCo: 2008.02. www.ieabioenergy.com

IEA Bioenergy (2013) Large industrial users of energy biomass. Task 40. Sustainable International Bioenergy Trade, International Energy Agency, Paris

IPCC (2013) Climate change 2013: the physical science basis. In: Contribution of working group I to the 5th assessment report of the intergovernmental panel on climate change, Cambridge University Press, Cambridge, and New York

Slade R, Saunders R, Gross R, Bauen A (2011) Energy from biomass: the size of the global resource. Imperial College Centre for Energy Policy and Technology and UK Energy Research Centre, London. ISBN: 1-903144-108

Smeets EMW, Faaij APC, Lewandowski IM, Turkenburg WC (2007) A bottom-up assessment and review of global bio-energy potentials to 2050. Prog Energy Combust Sci 33:56–106

Spliethoff H (2010) Power generation from solid fuels. Springer, Berlin, ISBN 978-3-642-02855-7

Stephenson AL, MacKay DJC (2014) Life cycle impacts of biomass electricity in 2020: scenarios for assessing the greenhouse gas impacts and energy input requirements of using North American woody biomass for electricity generation in the UK. Department of Energy and Climate Change, London

Stevens C, Brown RC (2011) Thermochemical processing of biomass: conversion into fuels, chemicals and power. Wiley, Chichester

Tilman DA, Amadeo JR, Kitto WD (1981) Wood combustion, principles, processes, and economics. Academic Press, New York

Webster AD (1919) Firewoods their production and fuel values. T Fisher Unwin Ltd., London

Williams A, Jones JM, Ma L, Pourkashanian M (2012) Pollutants from the combustion of biomass. Prog Energy Combust Sci 38:113–137

World Energy Outlook (2013) © OECD/IEA, 2013, Fig. 2.5, p 63

WHO (2014) Household air pollution and health. Fact sheet no. 292' updated Mar 2014. WHO, Geneva

Chapter 2
Combustion of Solid Biomass: Classification of Fuels

Abstract The combustion of solid biomass and the classification of these fuels are considered. Firstly the different methods of combustion appliances and plants are outlined from a fundamental point of view. The forms and types of solid biomass fuels, their chemical composition and the way they are classified are then described. Characterisation by chemical analysis and instrumental methods are outlined.

Keywords Combustion methods · Classification · Chemical analysis

2.1 Methods of Utilisation

Biomass is unique as a renewable energy source because it can be used for all three energy sectors electricity, heat and transport. This puts a high demand on the resources available. Solid biomass fuels have the advantage that a wide range of materials can be used. This flexibility also applies to the methods used for biomass utilisation which can vary depending on the nature of the output, heat or electricity and on the size of unit required. Domestic appliances up to ~25 kW include open fires, simple stoves, household heating boilers and cookers—including traditional cook stoves used widely in developing countries. The use of biomass for heating has recently been widely promoted, especially in the EU. Modern biomass furnaces incorporate state-of-the-art design, but there are still problems associated with meeting NO_x and particulate emissions limits.

Many of the methods used for the larger scale utilisation of solid biomass are based on technologies developed for conventional fuels such as coal and lignite. These methods are described in standard textbooks on combustion technology (Speight 2013; Spliethoff 2010). They use technologies such as fixed bed combustors (up to 5 MWth), moving or travelling grate combustors (up to 100 MWth), fluidised bed (up to 500 Me), suspension firing/pulverised fuel (pf) combustion and co-firing with fossil fuels (up to 900 MWe). Many industrial processes such as biomass co-firing

© The Author(s) 2014
J.M. Jones et al., *Pollutants Generated by the Combustion of Solid Biomass Fuels*,
SpringerBriefs in Applied Sciences and Technology,
DOI 10.1007/978-1-4471-6437-1_2

for power generation use only minor modifications to equipment designed for coal firing. This influences the pollutants formed.

Domestic heating is widely used throughout the world especially North America, Scandinavia, Europe and most Developing Countries. In Europe France is the largest consumer of wood for domestic heating, followed by Sweden, Finland and Germany. In the wintertime around half of all particulate organic matter in Paris has been attributed to biomass burning. Three methods available are open fires (up to 5 kWth), simple enclosed stoves, Patsari cook stoves and household heating boilers (3–10 kWth). Unlike larger units, these systems do not incorporate any flue gas after-treatment and as such emit significant particulate matter, unburned volatile organic matter and NO_x directly into the atmosphere. There is ongoing debate on the nature of these emissions from domestic applications as the rating of many units is based on steady state operation and does not take into account the initial start-up and refuelling emissions into account.

Pictures of typical small units are shown in Fig. 2.1, a typical African cook stove and an open domestic fire. The cook stove is insulated to improve its thermal efficiency.

Many design improvements aimed at improving the thermal efficiency of a system, such as improved air/fuel mixing and secondary air circulation will also reduce the pollutant emissions because the fuel will be burned more completely. But it is clear that the operation of a simple cooking unit or domestic stove will have a low efficiency and consequently high emissions from unburned fuel.

The basic features of a fixed bed combustor with fuel over-feed are shown in Fig. 2.2a. The main stages of combustion are shown but they do not occur in such a clearly defined sequential way in a real stove.

The fuel is added on top of the burning bed, consisting of logs, pieces of wood, briquettes, chips or wood pellets. The fuel undergoes devolatilisation forming carbon particles (containing ash), then these carbon particles react with CO_2 from the layer below producing CO, then the carbon particles falls to the oxidation zone forming CO_2 leaving ash. The ash, which will contain a significant amount of unburned carbon, falls through the grate and is removed periodically.

Fig. 2.1 Examples of domestic biomass combustion: **a** current technology cook stove, **b** domestic fire

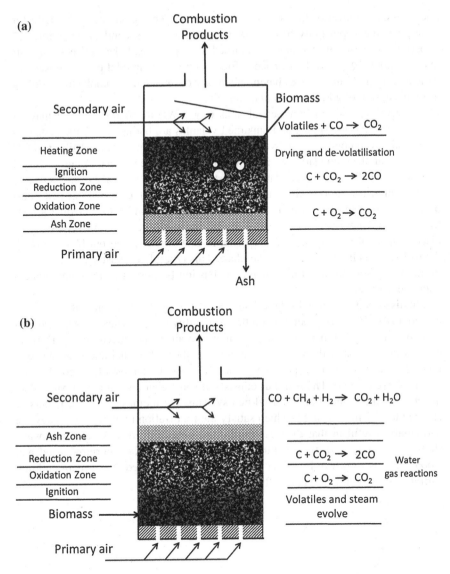

Fig. 2.2 a Idealised operation of a fixed bed combustor with over-feed arrangement. **b** Diagrammatic representation of an under-feed combustor

The problem with this type of combustor is that the smoke and unburned CO and volatiles escape from the top of the bed unless steps are taken to prevent that happening. In reality the movement of the particles leaves holes through the bed and so affects the air flow and hence the combustion process is not in a steady state. Most stoves in developed countries would have a deflector plate above the combustion region as shown in Fig. 2.2a so as to provide negative pressure in the flue to assist removal of the combustion products. The region above the deflector plate gives the opportunity for secondary combustion if secondary air is injected to permit the burn-out of soot and unburned hydrocarbons. However secondary air may be allowed to enter the furnace before that point by an inlet port or leakage as indicated in the figure. Stoves having a simpler design, such as cook stoves do not have this option of additional combustion air.

In larger domestic or commercial units (>30 kWt) a different method may be adopted in which the smoke and unburned volatiles are burned above the bed. This is the under-feed combustor shown in Fig. 2.2b. Here the unreacted volatiles and CO can react with the secondary air and their concentrations are reduced. Again such system would usually have a secondary combustion chamber above the bed and a flue.

Most industrial applications up to 100 MWth use either travelling bed combustion or fluidised bed combustion. Travelling grate combustors involve a continuously moving grate depicted in Fig. 2.3 but with the different stages taking place sequentially. A greater degree of burnout will take place but holes through the bed (called channelling) can take place. This type of combustor is widely used in incinerators as well as for biomass combustion because it can accommodate a wide range of feedstock sizes.

Fluidised beds are usually less than 100 MWth in size but some are built that can go up to 250 MWt depending on the availability of fuel. These systems consist of a combustion chamber into which combustion air is introduced via a perforated plate at the bottom with sufficient velocity that the bed of fuel and char above is fluidised. Because biomass produces about 90 % volatiles most of the combustion occurs above the bed. The bed thus consists of some char and an added particulate material such as sand. For fluidized-bed systems, the biomass fuels are pellets or chipped to 2–5 mm or larger which satisfy the fluidization requirements. They are particularly useful in that they can burn contaminated feedstock such as sewage sludge. A picture of a typical fluidised bed combustor plant burning wood and wood waste is shown in Fig. 2.4a. The combustor is situated in the main building surrounded by fuel handling and flue gas processing units.

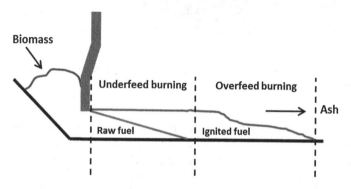

Fig. 2.3 Diagrammatic representation of a travelling bed combustor (not to scale—the length is shortened compared with a real unit)

Fig. 2.4 Examples of industrial applications **a** 44 MWe bubbling fluidised bed combustion facility using wood, Stevens Croft, Scotland (power technology), **b** Drax Power Station, UK (Picture by AFT)

The industrial combustion plants consuming the largest amount of biomass in the world at present are used for the combustion of waste from paper and pulp manufacture (although much of this is black liquor) and from agricultural waste such as bagasse and palm kernel (IEA 2010). However an increasing application is for power generation by co-firing biomass with coal or near 100 % biomass firing in conventional power stations. An example of a large power station is Drax Power Station in the UK (4 GWe total capacity) which is shown in Fig. 2.4a. Large quantities of solid biomass fuels are used by co-firing with coal in such power stations; this may involve blending a proportion of biomass with the pulverised coal ranging from 3 wt% up to nearly 100 % biomass. This method of biomass utilisation gives high generation efficiency, extensive emission control equipment and therefore reduced emissions per unit of electricity. Energy from biomass is predictable and controllable so can provide base load demand as well as peak demand if needed. Two of the units at Drax are currently being converted to 100 % biomass firing and the new biomass storage domes are shown in the bottom right side of Fig. 2.4b.

Pulverised fuel (pf) or suspended systems use biomass fuel particles that are reduced to a wide range of 10–1,000 μm or more in order to accomplish complete burnout in typically a furnace residence time of 5 s. The biomass fuel usually has to be within a specified range of properties to ensure good combustion and to minimise slagging and fouling and corrosion in the plant. The technique of using milled biomass started because of the ease of milling a small amount of biomass with coal and burning it in the same boilers designed for pulverised coal combustion. The same furnaces and burners with staged combustion are readily used for co-firing with little adjustment. A typical staged pulverised fuel burner is shown in Fig. 2.5. Devolatilisation takes place in the primary zone under fuel rich conditions and the resultant mixture of char, gaseous volatiles and soot are burned out with the secondary air. Normally these burners have to use three stages to minimise the formation of NO. If the burners are using 100 % biomass then their design has to be modified in order to get stable flames.

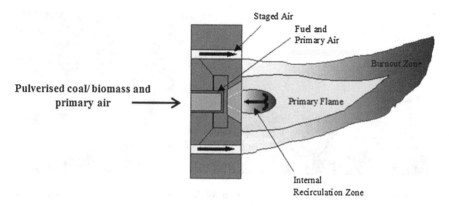

Fig. 2.5 Diagrammatic representation of a low-NO$_x$ staged burner

However most biomass does not mill as easily as coal because its fibrous nature results in poor grindability characteristics. This can result in larger particle sizes being produced, and increases wear on milling surfaces. This has implications for biomass handling methods. The effect on the times required for particle burn-out is discussed in Chap. 6. A comparison of a partially reacted coal particle and a partially reacted wood particle obtained using a scanning electron microscopy is shown in Fig. 2.6. The coal has melted so that the particle is nearly spherical and blow holes where the volatiles have escaped are dominant. In the case of the wood particle the remaining lignin is only partially melted and much of the original shape is retained at least initially.

Much of the handling technology for the larger units has been developed for coal or low grade fuels such as lignite and has now been adapted for biomass. There have been considerable efforts made to improve small units, but there are

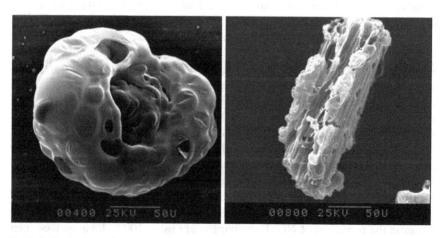

Fig. 2.6 Comparison of partially reacted fuel particle shapes (*left*) coal, (*right*) wood

inherent problems arising from their small size; these issues will be discussed later. Different biomass particle sizes are required for all these different systems, which impinge on the fuel supply and processing capabilities.

A further type of combustor that is available is the 'flameless' combustor where the flame temperature is about 1,000 °C and is not visible to the naked eye. This is a technique originally developed for the low NO_x combustion of gaseous fuels, but it has now been applied to the combustion of pulverised biomass (Abuelnuor et al. 2014).

2.2 Forms of Solid Biomass Fuels

The important issues are that there are multiple supply chains for solid biomass fuels, multiple applications, different scales of implementation, and a lack of an established infrastructure and standards. As a consequence it is not easy to generalise on the forms of solid biomass fuels on a global basis.

The UK Biomass Strategy identified the following biomass resources (DEFRA 2007)

- Conventional forestry
- Short rotation forestry
- Sawmill conversion products
- Agricultural crops and residues
- Oil bearing plants
- Animal products
- Municipal solid waste
- Industrial waste

The main sources of biomass are wood, energy crops and agricultural wastes. The most promising energy crops include Miscanthus, short rotation willow (SRW) and reed canary grass (RCG). Wastes include wheat straw, recycled waste wood in Europe and North America, bagasse, Palm Kernel and Olive wastes from the countries where these crops are grown.

The motivation for using biomass is to reduced greenhouse gas emissions, and so the sustainability and life cycle analysis of biomass supply should take into account emissions of all GHG including CO_2, N_2O and CH_4. Farming methods involving application of large amounts of fertiliser or clearing large land areas might increase net Greenhouse Warming Potential, GWP (Smith et al. 2012).

Some biomass materials such as agricultural residues may be used directly, but in the case of small appliances they first have to be made into pellets or briquettes. Solid biomass fuels are available in a variety of forms. These include logs and straw, and also processed products such as chips, pellets and pulverised fuels mainly from wood, straw and a range of agricultural residues. Logs are generally available in three standard sizes (25, 33, 50 cm), which fit the combustion chamber geometries of commercial log boilers. Similarly, wood chip standards [CEN/TS 14961: 2005 (E)] have been introduced in the EU to standardise properties such as dimension, moisture (range from dried to about 20 % moisture which can be stored, ash and nitrogen, and to provide information on heating value, density and chlorine content. In the case of pellets the standard [CEN/TS 14961: 2005 (E)] specifies the same properties: dimensions (based on diameter and length), moisture, ash, sulphur, mechanical durability, fines, additives, nitrogen. Information on heating value, bulk density and chlorine content are also recommended. There is a variety of chipping, grinding and pellet-making machines and, in the latter case, this often involves the capability for a certain degree of drying. In pellet making small particles (2–5 mm) of the biomass are compressed whilst steam heated and the lignin present acts as a binding agent although a binder can be added. Pellets are now an accepted form of fuel for many smaller units and since they can easily be transported and stored they are used as a fuel for pf power stations where the pellets are milled to provide the necessary particle size.

Direct firing of pulverised biomass or the co-firing with coal requires particles that are ground to less than 1 mm in size. Comminution of fibrous materials is difficult and expensive in terms of both energy and money because of the strength of the fibres present. Fibrous biomass requires about 5 times the energy of bituminous coal, however the process of torrefaction (van der Stelt et al. 2011; Basu 2013) can modify the nature of the products making it easier to mill. Torrefaction

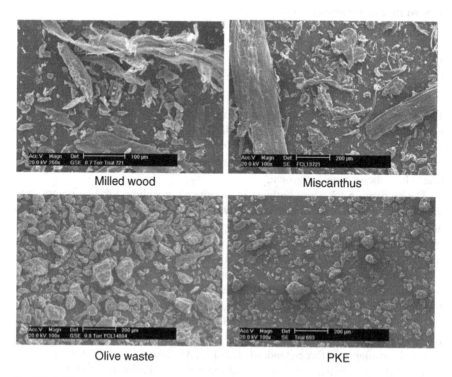

Fig. 2.7 Different biomass types illustrating the particle types resulting from milling

is the term used for roasting coffee and a similar process is used for biomass-it is heated at about 290 °C for about 30 min during which time much of the hemicellulose is lost, the degree of torrefaction being determined by the reaction conditions. Torrefied biomass has a higher energy density and if used for pf combustion grinds to smaller particles more readily. It also has advantages for use as a fuel for bed combustion. Agricultural process residues including nut products, such as Palm Kernel Expeller (PKE), and olive waste are available in ground form and are less fibrous than woods and grasses. The contrast between these different fuels the fibrous materials, wood (SRC, short rotation willow coppice) and straw, and the easily ground fuels, olive waste and PKE, is shown in the Fig. 2.7.

2.3 Types of Solid Biomass Fuels and Their Classification

As is the case with conventional fuels large data bases on the properties have been developed of which the most extensive is Phyllis 2 (2014). Proximate analysis is used to characterise biomass in order to measure its moisture, volatile matter, fixed carbon and ash contents. Ultimate analysis gives the actual chemical composition (usually C, H, N, S and O by difference). The metals content is also determined. Care has to be taken in the way in which these results are expressed and can be on as received basis (ar), dry basis (db), or dry ash free basis (daf). Values for the Higher Heating Value (HHV) are also obtained, this term is usually employed by the power generation industry but other terms used are calorific value or enthalpy of combustion. Details of these techniques are given in standard procedures which are available as CEN, BS, ASTM, NREL, Japanese and Chinese Standards. These methods of analysis were originally designed for use with coal and adapted for use with biomass, for example, by reducing the furnace temperature from 815 to 575 °C for the ash measurement, which avoids losses of volatile alkali metals. Some of these techniques developed for wood and straw are not necessarily applicable to some oriental biomass species with high protein content (Rabemanolontsoa et al. 2011).

Conventional solid fossil fuels are classified by a range of standard tests which are determined by a number of considerations. The commercial 'value' of a fuel as determined by the Higher Heating Value, volatiles, ash content and similar parameters is important, but so are safety considerations (flammability characteristics, biohazards etc.). Technically there are other criteria that determine the mode or ease of transportation or ease/efficiency of use in the end application (e.g. density, hardness, particle size). Finally, environmental factors (S, N contents) are also important. It should be noted that wood is a major feedstock for the paper industry and there are a wide range of standards for this industry which can sometimes be applicable to combustion. In particular the ratio of cellulose/lignin criterion is important in paper making, as are the methods of their measurement.

Until recently solid biomass fuels were not traded internationally on a large scale commercially and application was limited to the combustion of wood, logs, specialist applications of industrial waste from sugar cane bagasse, palm olive and from paper manufacture. Most of the applications were for domestic heating and cooking. The developments in the last decade of biomass on a large industrial scale have prompted the use of more rigorous standard test methods for not only combustion but also in handling and safety. These include safety in transportation and storage such as fire and biological hazards, the mechanical handling pellets or torrefied products and particularly friability and fine dust formation.

Difficulties arose came from the wide diversity of biomass feedstock, and a need exists for a comprehensive classification system which covers both physical and chemical specifications. The ultimate aim of such a system would be to allow the user to predict the behaviour of a biomass feedstock through its classification, preferably based on a few simple tests. Properties of biomass, which are important for combustion purposes, include pyrolysis behaviour, tar yield, volatile composition, together with the yield and composition of the char and its reactivity towards oxygen. Also, the inorganic composition is much more variable for biomass compared to coal, and this impacts on the combustion behaviour and needs to be considered. Clearly, measuring all these properties for each biomass feedstock is time-consuming and unrealistic; one aim is to consider the application of suitable correlations between them.

Biomass material can be broadly classified into groups based on the general assessment of their source. The major groups in this way shown here in Table 2.1 are (i) woody: pine chips, an energy crop, willow, (ii) herbaceous: two energy crops are listed—Miscanthus and Switchgrass, (iii) agricultural residues: wheat straw, rice husks, palm kernel expeller, bagasse (a sugar cane residue), olive residue or olive cake (the waste from olive oil mills), and animal wastes such as cow dung. Data for lignin and cellulose samples are also given in Table 2.1 for comparative purposes (the latter is an exact quantity but the value for lignin depends on the type of lignin). This broad classification is represented by examples in Table 2.1 together with analyses which are taken from the Energy Research Centre of the Netherlands (ECN) online data base, and information on a wide range of materials can be obtained from that compilation. It should be noted that the composition of many of these materials can vary depending on a number of factors, such as time of harvest, soil type, geographical location. The woods consist of two types, hardwoods and softwoods, the former are from deciduous trees (e.g. willow) and the latter from evergreens (e.g. pine). Typically the composition (wt%, dry basis) for hardwoods is: cellulose, 40–44 %, lignin 18–25 %; softwood, cellulose 40–44 % and lignin 25–35 %. The chemical nature of the hemicellulose and lignin are different. In hardwoods the hemicellulose consists of 10–35 wt% xylan whilst in softwoods the xylan content is 10–15 wt% and a major component is water soluble galactoglucoannan. Lignins consist of hydroxyphenylpropane units (described in the next chapter) and their content in hardwoods is 7–11 wt% and in softwoods it is 13 wt%. As already pointed out there is considerable variability of composition even in the same stem and between different trees.

Table 2.1 Typical proximate (db, %) and ultimate analyses % (daf) and higher heating values for a range of biomass types

Fuel (Oven dried)	Proximate analysis, db			Ultimate analysis, daf				HHV
	VM	FC	Ash	C	H	O	N	MJ/kg (dry)
Wood pine chips	80.0	19.4	0.6	52.1	6.1	41.44	0.3	21.02
Willow, SRC	83.4	15.01	1.59	51.0	6	42.9	0.1	18.86
Miscanthus giganteus	82.1	16.4	1.5	49.4	6.4	43.93	0.3	19.88
Switchgrass	82.94	14.38	2.69	47.52	5.96	44.8	0.37	19.07
Straw-wheat straw	71.89	17.8	10.32	46.35	6.29	47.92	4.04	18.18
Lignin	74.57	22.03	3.4	53.0	5.9	39.74	1.11	22.20
Cellulose	88.3	10.96	0.74	43.62	6.55	49.67	0.2	17.38

Source Phyllis 2 (2014)

The herbaceous biomass fuels and the agricultural fuels can be significantly different from the woods in terms of the trace metal and nitrogen contents and which have a significant bearing on the pollutant formation as described later.

2.4 Characterisation by Chemical Analysis

The data in Table 2.1 was obtained by chemical analysis which plays a major role in the understanding of the combustion behaviour of biomass. The major components are measured using the same analytical methods devised for coal although nitrogen and ash present a problem. Fuel-nitrogen can be difficult to measure because it can be in low concentrations necessitating the use of special methods. Ash can contain volatile components that prevent some of the older coal analysis methods being used as previously described.

All biomass contains the nutrients nitrogen, phosphorus and potassium and the other major essential elements are Ca, Mg, Na and Si, while minor species include Mn, Fe, Mo, Cu, and Zn. Chlorine is also present. A considerable amount of information is available about these species because they can have a very significant influence on slagging, fouling, corrosion in the furnace as well as pollutant formation. The single most important species is potassium. Straw would contain typically large amounts of potassium/chlorine: 0.2–0.97 dry% Cl and 0.7–1.3 % K; whilst woods would contain smaller amounts of potassium/chlorine: typically <0.1 % Cl and 0.2–0.5 % K. Silica tends to be a main component in the ash of herbaceous crops and residues (straws as well as rice husks), while calcium is higher in woods and some of the herbaceous energy crops. Detailed analyses of the C, O, H, N, S, Cl constituents have been given by Vassilev et al. (2010) and by Vassilev et al. (2013) for the inorganic constituents.

The simplest method of classification is based on chemical analysis. A method has been that suggested by Chen et al. (1998) which parallels the type of approach adopted previously for coal classification systems. This classification is based on

Fig. 2.8 Van Krevelen Plot of H/C against O/C for two types of biomass: woods *filled square*; herbaceous biomass, *triangle*. Values for lignin and cellulose are included (based on Jones et al. 2004)

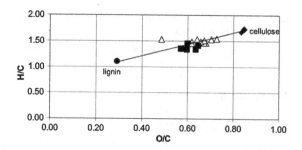

a van Krevelen diagram [plot of H/C versus O/C atomic ratios (daf)]. The premise in this classification is that biomass feed-stocks that fall within clusters in the van Krevelen diagram will have similar properties, regardless of their category (waste, wood, etc.). This is shown in Fig. 2.8 where only organic (non-aqueous) H and O is included (i.e. wt% H from ultimate analysis is corrected for H in moisture, and O is calculated by difference using daf data). This type of information can be correlated to the high heating values.

This type of plot has now been used by a number of research groups and can be used in the prediction of other properties, such as Higher Heating Value, and potentially in predicting lignin, and other similar quantities for lignocellulosic biomass. However some other parameters such as volatile matter or fixed carbon correlate less well, because these parameters are strongly influenced by both heating rate and ash content/composition. There is added difficulty in that the analyses vary across different particle size fractions of the heterogeneous biomass samples. Variation is seen between leaf and stem as well as bark and the analyses are affected by time of harvest, geographic origin, fertilizer treatment and length of storage.

The three main components, namely cellulose, hemicellulose and lignin can be used as a method of classification. In this approach the classification is based on the approximation that these major components behave independently during pyrolysis and combustion, and that the biomass behaviour can be predicted based on knowledge of the pure component behaviour. This type of classification is shown in Fig. 2.9.

The woody and herbaceous species are not identified individually in Fig. 2.8 since no particular pattern is present. What it does show is that the concentrations of cellulose and hemicellulose are on average about equal and that species in the top right hand corner would have low lignin concentrations. This may be used as an indicator of the tendency to form low levels of smoke and of high reactivity. However, because of the influence of the minerals present this cannot be used as a reliably at present. Potassium has a considerable effect on the pyrolysis mechanism and the influence of that would have to be included in any complete prediction scheme. The uncatalysed process favours the formation of sugar-type monomers from the carbohydrates, but the presence of potassium changes the mechanism to one that involves ring cracking to give gases and the polymerisation to char. It is known that the addition of potassium to biomass pyrolysis aids the production of charcoal.

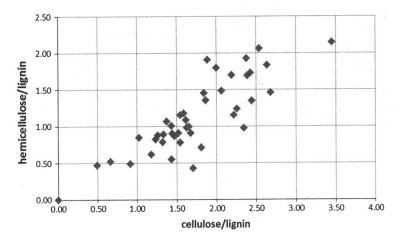

Fig. 2.9 Classification based on component composition for woody and herbaceous biomass samples (based on Jones et al. 2004)

2.5 Characterisation by TGA, PY-GC-MS and FTIR

Apart from chemical analysis it is possible to characterise a biomass by pyrolysis of a finely powdered sample in an inert gas stream or by combustion in air or oxygen. Laboratory scale thermogravimetric analysis (TGA) in air yields the burning profile, a mass loss trace. The first derivative of the weight loss curve gives characteristic temperatures for moisture loss, volatile loss, volatile ignition, char ignition and char burn-out. These methods were developed for coal but have been widely used for biomass in order to characterise the biomass in terms of ease of combustion and especially to determine the pyrolysis and char kinetics. This data can be obtained by slowly increasing the temperature or isothermally. Care has to be taken with these results because the upper temperature limit in a TGA is about 1,000 K which is below most char combustion conditions in many real combustion situations. If pyrolysis is undertaken in nitrogen or other inert gas with a heating rate of 5–40 K/s the mass loss curve produced shows the loss of water at about 100 °C and then a gradual loss due to pyrolysis only up to about a final temperature of 700 K when only carbon and ash remains and if oxygen is introduced then this can be burned out to leave ash. An example is shown in Fig. 2.10a. The derivative of this curve can be obtained to determine the kinetics and the difficulty is to decide at which point this is to be made.

In differential thermal analysis (DTA) the temperature differential between the sample and an inert reference is plotted giving a more sensitive indication of the reaction taking place. Figure 2.10b shows a differential thermal analysis of 'synthetic biomass', a mixture of cellulose (50 wt%), lignin (60 wt%), and xylan (20 wt%) which represents hemicellulose. In this plot a shoulder is visible on the main decomposition peak which is attributed to the hemicellulose decomposition,

Fig. 2.10 **a** An example of a pyrolysis TGA curve showing percentage mass loss against time. **b** An example of differential thermal analysis of a 'synthetic biomass', a mixture of cellulose, lignin, and xylan which represents hemicellulose (based on Nowakowski et al. 2007)

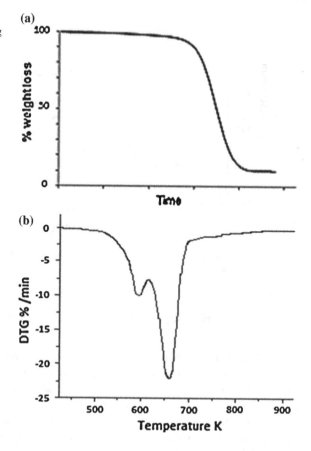

while the main peak is mainly attributed to cellulose decomposition. Lignin decomposition gives rise to the broad underlying peak. Although these features are over emphasised in this synthetic mixture it shows the same general features shown my most biomass samples such as woods wheat straw, reed canary grass and switchgrass.

Another method of characterising biomass types is illustrated in Fig. 2.11 and consists of a pyrolysis unit coupled to a gas chromatograph with the products analysed by a mass spectrometer (Py-GC-MS). This technique uses an electrically heated coil or filament to pyrolyse a sample at a known temperature so that the products can be analysed. Considerable advances have been made in the analysis of biomass products but an extremely important but difficult area is in the analysis of high molecular weight tars (Kandiyoti et al. 2006). In some cases FTIR can be used in place of the mass spectrometer and is particularly useful in the case of nitrogen compounds for example Darvell et al. (2012).

This type of characterisation of biomass is useful for several reasons including understanding the reactivity of a biomass but also the PAH and smoke forming tendency. The cellulose decomposition products consist mainly of oxygenated

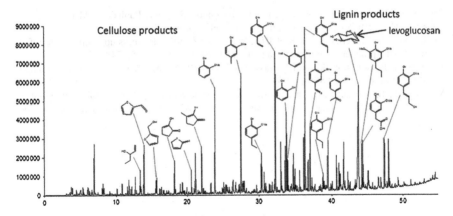

Fig. 2.11 Pyrolysis-GC-MS of pine wood with a pyrolysis temperature of 600 °C (*source* Dr. Emma Fitzpatrick)

products such as furfural and levoglucasan whilst the lignin components form largely aromatic compounds such as eugenol. The difference between pine, which is a softwood with high lignin content, and willow, which is a hardwood with a lower lignin content, is significant.

References

Abuelnuor AAA, Wahid MA, Hosseini SE, Saat A, Saqr KM, Sait HH, Osman M (2014) Characteristics of biomass in flameless combustion: a review. Renew Sustain Energy Rev 33(1):363–370

Basu P (2013) Biomass gasification, pyrolysis and torrefaction practical design and theory, 2nd edn. Elsevier Inc, Amsterdam

Chen Y, Carpenay S, Jenson A, Wojotowicz MA, Serio MA (1998) Modeling of biomass pyrolysis kinetics. Symp (Int) Combust 27:1327–1334

Darvell LI, Brindley C, Baxter XC, Jones JM, Williams A (2012) Nitrogen in biomass char and its fate during combustion—a model compound approach. Energy Fuels 26:6482–6491

DEFRA (2007) UK Biomass Strategy. DEFRA, London

Jones JM, Nawaz M, Darvell LI, Ross AB, Pourkashanian M, Williams A (2004) Towards biomass classification for energy applications, in science in thermal and chemical biomass conversion, Victoria, Canada, 30 Aug–2 Sept 2004

Kandiyoti R, Herod AA, Bartle KD (2006) Solid fuels and heavy hydrocarbon liquids. Elsevier, Oxford

Nowakowski DJ, Jones JM, Brydson RMD, Ross AB (2007) Potassium catalysis in the pyrolysis behaviour of short rotation willow coppice. Fuel 86:2389–2402

Phyllis 2 (2014) Database for biomass and waste. Energy Research Centre, Netherlands. https://www.ecn.nl/phyllis2

Rabemanolontsoa H, Ayada S, Saka S (2011) Quantitative method applicable for various biomass species to determine their chemical composition. Biomass Bioenergy 35:4630–4635

Smith KA, Mosier AR, Crutzen PJ, Winiwarter W (2012) The role of N_2O derived from crop-based biofuels, and from agriculture in general, in Earth's climate. Philos Trans R Soc B 367:1169–1174

Speight JG (2013) Coal-fired power generation handbook. Scrivener Publishing, MA

Spliethoff H (2010) Power generation from solid fuels. Springer, London

van der Stelt MJC, Gerhauser H, Kiel JHA, Ptasinski KJ (2011) Biomass upgrading by torrefaction for the production of biofuels: a review. Biomass Bioenergy 35:3748–3762

Vassilev SV, Baxter D, Andersen LK, Vassileva CG (2010) An overview of the chemical composition of biomass. Fuel 89:913–933

Vassilev SV, Baxter D, Vassileva CG (2013) An overview of the behaviour of biomass during combustion: Part I. Phase-mineral transformations of organic and inorganic matter. Fuel 112:391–449

Chapter 3
The Combustion of Solid Biomass

Abstract The combustion of solid biomass is covered in this chapter. This covers the general mechanism of combustion, moisture evaporation, devolatilisation, the combustion of the volatiles gases and tars and finally char combustion. Details are given about the kinetics of these reactions, the devolatilisation products which if unburned give CO and organic products and char combustion which if incompletely burned result in fine char particles in the combustion products.

Keywords Devolatilisation · Volatile products · Char combustion

3.1 General Mechanism of Combustion

There are several ways in which solid biomass can be utilised. It can be burned directly in a combustion chamber or it can be used indirectly via the process of gasification. In the latter process the products are gasified and these gaseous products then burned in a boiler or furnace or in an engine such as a gas turbine or a reciprocating engine. Direct combustion is the major route used at present and so little attention will be given here to gasification of biomass however the technology is covered elsewhere (Higman and van der Burgt 2008; Basu 2013). Gasification is associated with two technical difficulties, firstly the gaseous product obtained has a low HHV and secondly the products contain significant amounts of tar which is difficult to remove (Higman and van der Burgt 2008). For economic reasons gasification is most easily used with a low cost feedstock such as domestic refuse. Another possible method of utilisation uses finely pulverised solid biomass which is injected directly into an engine such as a gas turbine; but this is still at an early research stage (Piriou et al. 2013).

When a small solid particle of a carbonaceous material such as biomass enters a hot oxidising atmosphere it will ignite and burn in a manner shown in Fig. 3.1. This is the simplest example of biomass combustion and is the one followed in a power station boiler. It is analogous to the combustion of pulverised coal which has been extensively studied. The combustion consists of the following stages (a) heating-up

© The Author(s) 2014

J.M. Jones et al., *Pollutants Generated by the Combustion of Solid Biomass Fuels*,
SpringerBriefs in Applied Sciences and Technology,
DOI 10.1007/978-1-4471-6437-1_3

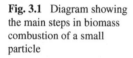

Fig. 3.1 Diagram showing the main steps in biomass combustion of a small particle

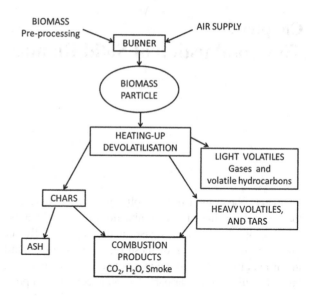

to a temperature where evaporation of any moisture or volatile oils takes place; decomposition of the carbonaceous structure termed 'devolatilisation' to produce volatile gases which consist of tars and gases, and a char. These volatile products will burn as a diffusion flame around the particle and then as the amount of volatile products formed declines oxygen penetrates to the char surface and heterogeneous combustion of the char takes place. The exact process that takes place is dependent on the type of the biomass, the particle size and the type of combustor. In large particles such as in grate combustion the combustion processes may occur sequentially as a thermal wave through the particle. Combustion of very large biomass objects has been extensively studied in relation to research into fires in buildings but the conditions are significantly different.

For combustion to take place there must be sufficient air in the proximity of the fuel particle to give at least a stoichiometric mixture but usually there is excess air. This is because the inflow of air is not perfectly directed to the fuel nor is the mixing process ideal. If ø is the mixture ratio (in mol fraction), then if ø = 1 it is stoichiometric and if it is greater than 1 it is fuel lean and there is excess air (or oxygen) in the combustion products. If it is less than I then combustion would be incomplete and gasification takes place. The combustion steps are

Fuel + ø air → {dissociated products in the combustor at the flame temperature}

→ [cooling in the flue] → {final combustion products}

During cooling in the exhaust system the dissociated species are quenched to give the final gaseous and particulate products.

The overall chemical reaction giving the final combustion products is:

$$C, H, O, N, S, \text{(metals)} + O_2 = CO_2 + H_2O + \text{metal oxides,}$$

chlorides and sulphates.

The composition of the products can be calculated from simple stoichiometry and an example is given in Appendix A. The composition of the dissociated products in the flame reactor at a known temperature or flame temperatures can be calculated by computer programs such as Chemkin (2014) and many other 'equilibrium' programs that are online.

The basic steps of the chemical mechanism are as follows

$$\text{Wet Biomass heating up/drying} = \text{dry biomass} \tag{R 3.1}$$

$$\text{Biomass} = \text{volatiles (tars and gases)} + \text{char} \tag{R 3.2}$$

$$\text{Volatiles} + \text{air} = CO + CO_2 + (\text{PAH, unburned hydrocarbons,}$$
$$\text{soot, inorganic aerosols}) \tag{R 3.3}$$

$$\text{Char} + \text{air} = CO + CO_2 \tag{R 3.4}$$

$$\text{Volatiles(N, S, K, etc.)} = \text{N, S, K based pollutants} \tag{R 3.5}$$

$$\text{Char(N, S, K, etc.)} = \text{N, S, K based pollutants} \tag{R 3.6}$$

Early models of biomass combustion used simplified rate expressions for each of these steps, while recent models include more detailed, biomass-specific chemistry to better predict burn-out and pollution formation (Williams et al. 2012). CFD modelling of these steps is discussed in Chap. 6. Pollutants are formed alongside the main combustion reactions from the N, S, Cl, K as well as other trace elements contained in the volatiles and char. If the combustion is incomplete, due to factors such as local stoichiometry (poor mixing), temperature, residence time etc. then products such as CO, polyaromatic hydrocarbons (PAH) and soot, together with characteristic biotracers of biomass combustion such as levoglucosan, guaiacols, phytosterols and substituted syringols are released. Thus, the atmospheric emissions can contain tar aerosols and soot, which with fine char particles and metal-based aerosols such as KCl all together form smoke. The nitrogen compounds are partially released with the volatiles, whilst some forms a C–N matrix in the char which is then released during the char combustion stage forming NO_x and the NO_x precursors, HCN and NH_3. Sulphur is released as SO_2 during both volatile and char combustion. Other metal containing compounds such KCl, KOH together the sulphur compounds form a range of gas phase species, which can be released as aerosols, but importantly also deposit in combustion chambers.

The development of models to predict the formation of these pollutants has been a continuing research effort. These have been applied to a range of applications and particle sizes, ranging from small pulverised particles less than 1 mm in diameter, to chip, or larger still such as logs, and at the extreme to trees in forest fires. Because of the coupling between chemistry and heat and mass transfer during particle conversion, fuel size has a major effect on the emissions. Although the general features are well understood, there is still debate regarding the relative roles of the controlling mechanisms. Pyrolysis might be controlled by either the chemical kinetic rate or the internal heat transport rate, and char oxidation might be controlled by the chemical kinetic rate or internal or external diffusion rates. The actual rates in the regimes for drying, pyrolysis, char formation and char combustion depend on particle size and can be defined as thermally thin, thermally thick with thermal wave regimes. In the first case, the temperature is constant across the particle and this is normally assumed in the heating up step for small (pf) particles in the drying and pyrolysis steps. For larger particles, in fluidised or fixed beds, a thermal wave is assumed to pass through the particle, causing sequential drying, pyrolysis, char formation and then combustion. In this situation all the regimes coexist. The thermally thick case is assumed when there is large biomass particle involved and considerable thermal gradients.

3.2 Particle Heating and Moisture Evaporation

Particles of biomass may contain moisture and oils which on heating undergo evaporation following the laws given later in Eq. 3.1. This heating process is determined by the surrounding temperature and the size of the particle. The extent of the temperature uniformity is defined by the Biot number and the application of this concept can be applied in a number of ways; here the method set out by Hayhurst (2013) is adopted. The Biot number is the ratio of the two characteristic times for a fuel particle to be heated up to the temperature of its surroundings, when that heating is controlled, respectively, by internal and external heat transfer. Thus $Bi = \{t_{heatint}/t_{heatext}\}$. Assuming that the heating which is a measure of the importance of internal temperature gradients. For large values of Bi, the internal heat transfer is slow compared to the external heat transfer and internal temperature gradients are significant. This leads to the idea of a decomposition temperature, at which, $t_{kinetics}$ equals either $t_{heatint}$ (for Bi > 1) or $t_{heatext}$ (when Bi < 1), with $t_{kinetics}$ characterised by the particle's temperature and both $t_{heatint}$ and $t_{heatext}$ defined by its size, thermal properties etc. In that case, if Bi > 1, reaction fronts move into the particle, creating a "shrinking core" of material, where the reaction in question has not yet occurred. On the other hand, if Bi < 1, that reaction always proceeds uniformly throughout the entire particle, and whenever its decomposition temperature is reached, there is a change from kinetic control of that step to external heat transfer. In this Chapter we assume the particle is uniformly heated and the other cases with larger particles, such as fixed beds are considered in Chap. 6.

The moisture content in biomass plays a significant, sometimes very significant role in the combustion process. Typically newly cut woods contain up to about 50 wt% moisture, present as bound (about 30 wt%) and free water in the pores and capillaries. Even after considerable ambient drying, the moisture content is still about 10–20 %. For some materials such as bagasse the moisture content is kept at about 50 % and it is not dried to prevent spontaneous combustion, and is burned in the furnace still containing that amount of moisture.

Usually the biomass fuel is dried to about 10 % moisture to aid handling, transportation and storage. For combustion of 100 % pulverised biomass (suspension firing) of small particles or in co-firing with pf coal the particles heat up rapidly and moisture lost early in the furnace, but there is an efficiency loss due to the latent energy of water evaporation from the biomass. Because of this drying process may be carried separately especially if the fuel is converted to pellets or torrefied, and this significantly affects the temperature and stability of the flame. However in the large-scale utilisation of biomass the drying process is energy intensive and an optimum situation has to be decided for any particular application.

Many drying models have been developed, and early models for biomass drying were those used for pulverised coal (Peters 2003; Lu et al. 2010), but recently, more detailed models have been developed for larger particles. For example, in the case of a small particle it is assumed that the progress of drying is limited by the transport of heat into the particle, and the moisture evaporation rate, $\dot{r}_M$, can be approximated (Peters 2003) by Eq. (3.1):

$$
\begin{aligned}
&\text{If } T_s \geq T_{\text{evap}} \quad \dot{r}_M = f_M \frac{(T_s - T_{\text{evap}})\rho_M c_{pM}}{\Delta H_M \delta t} \\
&\text{if } T_s < T_{\text{evap}} \quad \dot{r}_M = 0
\end{aligned}
\tag{3.1}
$$

where T_s is the surrounding temperature, $f_M = X_M/M$. In the case of the drying of small particles, most computer models, such as that by ANSYS Fluent assume that there is a uniform temperature profiles across the particle. The use of non-uniform temperature profiles is necessary for bed firing and the use of larger particles in fluidized bed and some pf studies as described later.

3.3 Devolatilisation

Devolatilisation of initial solid biomass particles occurs in the early stages of combustion when they enter the flame zone and reach a sufficiently high temperature for rapid chemical decomposition or pyrolysis to take place. A large part of the original biomass is converted to gases and tars, the so-called 'volatiles' which consists of inorganic species such as CO, CO_2, H_2O, gases such as methane and a range of higher hydrocarbons. A small amount of residual biomass char (about 30–5 wt%) is formed, the amount of conversion depending on the heating rate and final temperature of the particle. The term used for biomass char can vary from charcoal to biochar or just char depending on the mode of formation. Charcoal is the name used for large quantities of material produced by a slow heating process and is used as a fuel usually in stoves. The term 'Biochar' is normally used for char produced for soil remedial purposes and which will also undertake some carbon dioxide sequestration.

From an environmental pollution point of view the issue is whether the volatile gases and tars burn completely out in the combustion chamber or pass through it without burning completely and are vented into the surrounding air. If they do then this is a source of volatile organic compounds which would include a substantial amount of methane.

Whilst the devolatilisation behaviour of most major coals has been studied for over a hundred years and is reasonably well known there is considerable uncertainty about biomass devolatilisation rates, especially at high temperature. It is noted that, like coal, the rate and the total yield of volatiles are strongly influenced by the conditions under which the process occurs, in particular by the temperature and heating rate of the particle. Typically the temperature at which pyrolysis starts for biomass is about 160–250 °C compared to about 350 °C for a typical bituminous coal, and moisture and oils will be lost earlier causing problems in the identification of true pyrolysis rates. Data obtained using traditional methods, such as drop tubes, heated wire mesh and fluidized bed reactors at temperatures of 800–1,200 °C are suitable for some applications but do not extrapolate well to the higher temperature that exist in a pf combustion furnace.

It has been found that for both coal and biomass the quantity of volatiles released under high heating rates is greater than the proximate value determined by the laboratory test method. In addition, the different procedures that are used to derive the devolatilisation kinetics result in considerable diversity in the literature. This arises from many factors including differences in experimental technique, uncertainty of the exact composition of a particular biomass, the effects of secondary reactions and the catalytic influence of reactive metals such as potassium present in a sample.

As described in Sect. 2.4, solid biomass consists of three major components of cellulose, hemicellulose and lignin linked together as well easily extractable/ volatile oils as well as inorganic components. The exact quantities of these species depend on the type of biomass and can be determined analytically, and they determine the processes that would take place during combustion from both a physical and chemical point of view.

Cellulose is a crystalline polymer of glucose ($-C_6H_{10}O_5-$) and the basic unit is illustrated in Fig. 3.2a. Hemicellulose is a complex mixed amorphous polymer of 5- and 6-carbon sugars and is the most reactive component in biomass for that reason. Lignin is a random three-dimensional phenolic polymer of which there are three classes based on different phenyl-propane monomeric units. These are guiacyl (G), p-hydroxy-phenol (H) and syringyl (S) and these are shown in Fig. 3.2b. Their respective chemical formulae are $C_9H_{10}O_2$, $C_{10}H_{12}O_3$ and $C_{11}H_{14}O_4$. Because of the increasing number of methoxy groups the C/H and C/O ratios are significantly different; the C/H ratios are 0.90, 0.83 and 0.78 respectively whilst the C/O ratios are 4.5, 3.3 and 2.75 respectively. The different biomass fuels contain different amounts of G:H:S. In general softwoods are based on G units, hardwoods are based on G/S and grasses on H units. Because of the differing C/H and C/O ratios this implies not only difference in HHV but also in the tendency to form PAH and smoke.

Biomass may be considered to consist of the three major components (cellulose, hemicellulose and lignin) all of which consist of subunits and all of which are

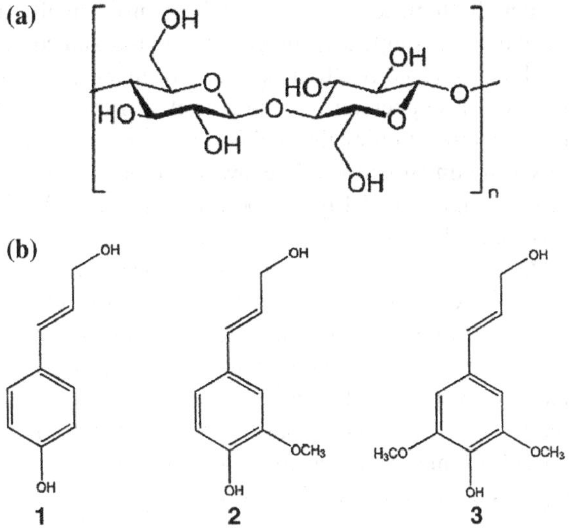

Fig. 3.2 Chemical structures for **a** cellulose unit, **b** for some lignin monomers: (*1*) paracoumaryl alcohol, (*2*) coniferyl alcohol and (*3*) sinapyl alcohol

chemically linked, as well as loosely bound extractives and water. Consequently the devolatilisation can only be fully described by a large number of different sequential and parallel reaction steps and thus simple pyrolysis models have been assumed. In the case of coal this was achieved most commonly by using one overall reaction, or two competing reactions or a distributed reaction activation energy approach. These methods have been used with biomass too but an approach which is useful for biomass combustion generally is to use the fact that biomass consists of three major components which are assumed to react independently. For example, pine wood typically contains 45 % cellulose, 29 % lignin and 26 % hemicellulose, and reaction paths and rates would be assumed for the three components. This approach has to be modified for the particular application and there two major groups thus: (i) the high temperature, high heating rate rapid devolatilisation during the combustion of a pulverised fuel in, for example, a power station boiler, and (ii) a biomass combustion fixed bed unit would provide the other extreme. Fluidised bed combustion would probably fall into the first category depending on the conditions.

During pyrolysis the hemicellulose compounds react first, then the cellulose compounds followed by the lignocellulose/lignin complex. These decompose to cyclic hydrocarbons, acids and then form CO_2, CO and H_2. The lignin macromolecules decompose over the whole temperature range giving variants of the monomers shown in Fig. 3.1b but depending on the temperature, and of course the type of biomass. If TGA experiments are used to determine the activation energy then the early part of the decomposition curve is used to deduce the activation energy and this gives the values for the hemicellulose decomposition. The higher temperature range will give information for the cellulose and lignin decomposition region (Branca and Di Blasi 2013; Lasode et al. 2014). In the case of pure compounds such as cellulose then a single activation energy can be obtained to approximate its decomposition (for example, Saddawi et al. 2010).

For the high temperature combustion of small particles in, for example a pf combustor, a single step process is often assumed and this is acceptable because of the very rapid reaction rates where much of the reaction is controlled by the increase in temperature of the particle. In this case the reaction is dominated by the decomposition of the main component which is cellulose decomposition and linked lignin subgroups (Saddawi et al. 2010). For lower temperature situations a more complex model has to be used (Hayhurst 2013; Branca and Di Blasi 2013; Lasode et al. 2014).

Many measurements have been made of the devolatilisation rates of biomass using TGA methods and interpretation in terms of kinetic factors. As pointed out it is difficult to extract exact kinetic parameters because of the changing reaction but also because of issues relating to the heating rate of the particles and the interpretation of results (Weber 2008; Hayhurst 2013; Pickard et al. 2013). The simplest model is based on an overall single kinetic step. In this situation, the rate of volatile release may be assumed to be first-order dependent on the amount of volatiles remaining in the particle as follows,

$$dV/dt = k[1 - V] \tag{3.2}$$

$$\text{where } k = A \exp(-E/RT_p) \tag{3.3}$$

where the rate constant, k, is expressed in the Arrhenius form in Eq. 3.3 in which Tp is the temperature of the particle, and A and E are the pre-exponential factor and the activation energy, respectively, which need to be determined experimentally for a particular solid fuel.

There a considerable amount of data available in the literature obtained using TGA at temperatures up to about 1,000 K but little available obtained using high temperature environments because of the rapid rates of reaction. Information can be derived from the Network Models such as the FG-BioMass model (Chen et al. 1998), bio-FLASHCHAIN (Niksa 2000) and Bio-CPD (Sheng and Azevedo 2002). The functional group approach can yield multiple simultaneous first-order rates for various products, but their interpretation of these can be difficult. An added difficulty comes from the fact that the potassium concentration in the biomass has some influence on devolatilisation (Jones et al. 2010; Saddawi et al. 2012) and so washing biomass changes the rate of devolatilisation. It was found that the amount of tar produced during combustion is changed by removal of the potassium by washing. A number of studies have been made on the rate of release of potassium from biomass combustion (Jensen et al. 2000; van Lith et al. 2008; Jones et al. 2010) and about 30 % of this occurs during devolatilisation.

If details of the devolatilisation of the fuel are known then a multiple reaction rate model may be employed. In the case of coal a two competing rates model (Kobayashi et al. 1977) has been used, but with biomass it is more appropriate to use a three reaction rate model, the reactions applying to the decomposition of the three major biomass components, hemicellulose, cellulose and lignin. This would be:

$$k_1 = A_1 e^{-(E_1/RT_p)}, \quad k_2 = A_2 e^{-(E_2/RT_p)}, \quad k_3 = A_3 e^{-(E_3/RT_p)} \tag{3.4}$$

where k_1, k_2 and k_3 represent the reaction rate constants for each component. The difficulty with this is in acquiring these values for any particular biomass and there are advances here in relation to this by the provision of some specific data by Branca and Di Blasi (2013) and Lasode et al. (2014).

Figure 3.3 shows kinetic data for some of the reactions. For high temperature pf combustion studies high activation energy values were chosen by Ma et al. (2007) based on evidence set out by Jones et al. (1999). For pine wood the values used are $A = 6 \times 10^{13}$ $1s^{-1}$ and $E = 250$ kJ/mol and this is shown in Fig. 3.3 as the High Temperature expression. This data was used in a model in conjunction with the heating up equations for larger particles. An alternative approach is to couple the heating-up equations with the kinetics so that one set of apparent kinetic parameter provides all the relevant data (Blondeau and Jeanmart 2011). The region marked flame arises from one data point derived from the lifetime of a very small wood particle travelling through a flame at 1,500 K. However from Fig. 3.3 it is seen that half-life ($t_{1/2} = \ln2/k$) is about 1 ms and consequently that particle would have reacted before flame temperature is reached, i.e. the point should be further to the right. It also means that in pf flames the particle temperature is the dominating factor not the kinetic rate expression.

The cellulose line in Fig. 3.3 is from the study by Saddawi et al. (2010) where $E = 178.7$ kJ/mol. The area termed Mixed Biomass comes from a number of studies in the literature for woods and agricultural residues including those of Jones et al. (2007), Branca and Di Blasi (2013) and Lasode et al. (2014). Recently Branca and Di Blasi (2013) have determined activation energies for devolatilisation of hardwood and soft wood of 200 kJ/mol for cellulose and 176 kJ/mol for lignin. The values obtained for hemicellulose for hardwood and softwood were 147 and 110 kJ/mol respectively. This data are consistent and that of Lasode et al. (2014). Their values for cellulose and lignin are approximately similar to

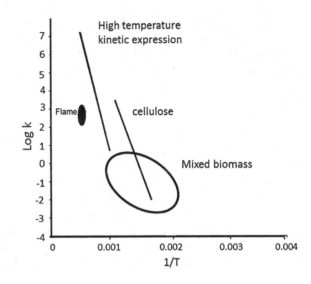

Fig. 3.3 Comparison of kinetic rates for biomass-plot of $\log_{10}$ (s^{-1}) against 1/T (K) for thermal decomposition of biomass where *solid lines* represent pure species and the *oval* highlights the curves generated by work using a number of different biomass fuels and includes demineralised willow

the cellulose curve in Fig. 3.3 and the hemicellulose values similar to the Mixed Biomass region.

For most applications it seems that a two competing model is satisfactory and for pf combustion at high temperatures a single reaction rate is adequate.

3.4 Combustion of the Volatiles—Gases and Tars

The gases and volatiles released make up 70–95 % of the devolatilisation products depending on the heating rate and final temperature, and much of this is CO and H_2. As an extreme example, finely divided wood or straw burns with a blue flame when sprinkled into a methane flame, which is in marked contrast to finely divided coal which burns with a yellow flame due to the soot and ash content. Larger particles do however form larger amounts of char which then has a yellow flame. Obviously the combustion of these gases and tars make up a significant part of the heat of combustion of biomass. The composition and hence calorific value of the gases, tars and chars derived from biomass depend on the heating rate and final temperature. Unlike coal char the generated biomass char contains a significant amount of oxygen the amount of which is determined by the final temperature. The oxygen content decreases with an increase in temperature under pyrolysis conditions but there is a shortage of information obtained directly from combustion situations. There is a vast amount of information on the composition of the devolatilisation gases produced and details are given by for example Di Blasi (2008), Klass (1998) and Blondeau and Jeanmart (2011).

Activation energies can be obtained using the Network pyrolysis programs previously described, that is, FG-BioMass, Bio-CPD and bio-FLASHCHAIN. These are derived from earlier coal pyrolysis models and can give information on the gas species, tars and the composition of the char. Additional information can be obtained on the nitrogen partition between the char and the other species which is useful information for NO_x prediction models. In the simplest models it is sufficient to follow the reaction steps (R 3.2) and (R 3.3), namely that the biomass produces volatiles which burn to form CO and then CO_2. The composition of the gases, tars and char from pine wood pyrolysis calculated by FG-BioMass for a range of temperatures shown in Fig. 3.4. Input from such calculations can be used for detailed combustion models.

It is seen that this calculation predicts that the char yield decreases as the temperature increases and the tar yield increases. Experimental evidence suggests that at higher temperatures and higher heating rates the char yield is about 90 wt%. One problem is how to handle the tars produced in a combustion situation. One option is to assume that the tar is an intermediate product and undergoes further secondary reaction by combustion or by pyrolysis or just escapes from the reaction zone. This last option occurs in gasifiers and in part in fixed bed reactors. If the tar is treated as an intermediate product the composition can be determined either by chemical analysis or using a computer code such as FG-Biomass. Typical

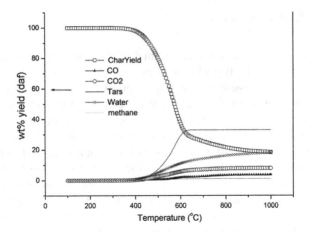

Fig. 3.4 Pyrolysis products calculated at different temperature using FG-BioMass

tar compositions are approximately $C_6H_6O_{0.3}$ but this depends on the degree of pyrolysis. In high temperature furnaces the tar is rapidly converted to gases and soot. At high temperatures little tar is formed by pyrolysis, little above 1,000 °C and none above 1,200 °C. Consequently the importance of tar formation depends on the application.

More detailed combustion models have been developed to calculate the formation of pyrolysis gases from biomass (Faravelli et al. 2013; Ranzi et al. 2014). This model gives the steps for the decomposition of biomass followed by reaction steps leading to the devolatilisation products which was developed for flash pyrolysis. With some adaptation it could be used in a combustion model. An abridged version is set out in Fig. 3.5.

In this model biomass is considered to consist of the three species cellulose (CELL), hemicellose (HSE) and lignin (LIG), which is considered to consist of three structural types in order to represent the different types of hard and soft wood. Cellulose is characterized by the glucose monomer ($-C_6H_{10}O_5-$) and in the case hemicellulose, the basic component is considered to be xylan ($-C_5H_8O_4-$). Lignins are represented by a combination of three monomeric units: LIGC ($C_{17}H_{17}O_5$) which is rich in carbon; LIGH ($C_{22}H_{29}O_9$) which is rich in hydrogen, and LIGO ($C_{20}H_{23}O_{10}$) which is rich in oxygen.

CELL $\longrightarrow$ CELLA, H_2O and Char
CELLA $\longrightarrow$ Gases, Liquids and Char
HCE $\longrightarrow$ 0.4 HCE1 + 0.6 HCE2
HCE1 $\longrightarrow$ XYLAN
HCE1, HCE2 $\longrightarrow$ Gases, Liquids and Char
LIG-C, LIG-H, LIG-O $\longrightarrow$ Gases, High Molecular Weight Products and Char

Fig. 3.5 Multi-step kinetic model of biomass devolatilisation (Faravelli et al. 2013)

3.5 Char Combustion

The biomass char combustion mechanism is similar to that developed for coal char combustion, and the typical type of model is set out in Fig. 3.6. The formation of char and its combustion is an important process. It is partially responsible for the formation of NO and CO in the flue gases and it determines the physical and chemical structure of the char in the combustor which in turn determines the fragmentation of the char resulting in the emission of the carbonaceous particulate matter from the combustor. These particles become entrained in the flue gas flow and whilst larger particles fall out into the chimney or removed by abatement equipment, fine particles are emitted into the atmosphere. Larger scale industrial plant with higher flow rates of flue gases may entrain a greater amount of these larger particles but these should be removed by cyclones or similar devices. A second issue relating to char burning rates is that of the unburned-carbon-in-ash. In many grate based units this can be very significant, often 10–30 wt% but it can be significant in pulverized fuel biomass plants and can be up to 10 wt% of the ash. Finally the distribution of fuel-N and metals particularly potassium between the char and the volatiles is essential information for modelling purposes. The bio-char combustion models follow those typically developed for coal char combustion and a general model is set out in Fig. 3.6.

In the case of coal char combustion information is available on intrinsic reactivity and annealing effects which are available in self-contained models e.g. Hurt et al. (1998). At high temperatures, chemisorption of gaseous species to carbon surfaces results in carbon-oxygen surface complexes such as carbonyl groups (Haynes and Newbury 2000). Catalytic elements among the alkaline and alkaline earth metals strongly influence the reactions between carbon and O_2, CO_2 or H_2O. Due to the heterogeneity characterizing nearly all carbon surfaces, particularly non-homogeneous char-C surfaces, and to variations in carbon-oxygen complexes and catalytic elements, there is an activation energy distribution for adsorption and

Fig. 3.6 Outline of the biomass char combustion steps

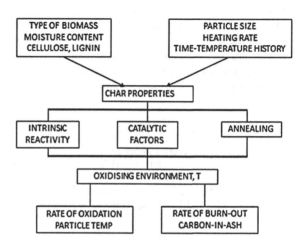

desorption. Char structures and activation energy distributions are fuel specific and depends on the initial volatile content of the parent fuel, initial porosity, heating rate and gas atmosphere during devolatilization and mineral matter content etc.

The concentrations and number of active sites varies as a function of carbon conversion. For example active sites associated with low activation energies can be consumed more rapidly than sites associated with high activation energies. As a result, the activation energy distribution can change as a function of conversion. This complex reaction scheme can be simplified and at high temperature, say above 600 °C, when the reaction with oxygen includes the reactions:

$$2C_f + O_2 \rightarrow C(O) + C(O) \tag{R 3.7}$$

$$C(O) \rightarrow CO + C_f \tag{R 3.8}$$

$$C(O) + C(O) \rightarrow C_f + CO_2 \tag{R 3.9}$$

But biomass char contains oxygen the amount being about 30 % at 1,200 K and decreases to close to zero at 1,800 K. This significant point has not been completely assimilated in current biomass char reaction models (Backreedy et al. 2002).

Char burnout depends on a number of parameters such as the temperature, char reactivity, particle surface area, oxygen concentration, residence time, ash inhibition and catalytic effects. The char burning rate, R, can be generally expressed as:

$$R = f\left(A_p, T_p, \rho_p, p_{ox}, k_{eff}\right) \tag{3.5}$$

Here the intrinsic reaction rate is used but usually it is simplified in terms of the apparent reaction rate where the surface area is included in the term. This is because this is a much easier quantity to determine experimentally. In pulverised biomass combustion (that is, with small char particles), the models assume that the char particle is surrounded by a stagnant gas boundary-layer through which the oxygen must diffuse before it reacts with the char surface. Whilst a continuous layer should be used the assumption of a single-film model for reactant penetration and oxidation and gasification reactions if necessary is sufficiently accurate for most situations including oxy-char combustion (Hecht et al. 2013).

Therefore, the overall oxidation rate of the char may be limited by either the diffusion rate of oxygen to the external surface of the char particle or by the effective char reactivity itself, or by the combination of the two. However, at a high temperature, diffusion is often the limiting process of the char combustion. Mathematically, this is usually expressed as an implicit function as follows:

$$R = R_c \left(C_g - \frac{R}{D_0}\right)^N \tag{3.6}$$

where D_0 is the bulk diffusion coefficient, C_g means reacting gas species concentration in the bulk, R_c is the chemical kinetic rate and N is known as the dimensionless particle reaction order (Smith 1987) $N = 0$ and 1 are the two typical reaction orders that have been used in the many combustion CFD modelling applications. A zero order of reaction, i.e. $N = 0$ represent a diffusion control surface reaction. In this situation, the knowledge of the bulk diffusivity for the oxidant is the only requirement.

The intrinsic char combustion rate (dm_p/dt) for a global reaction order of unity $(N = 1)$ can be expressed as

$$\frac{dm_p}{dt} = -A_p p_{ox} \frac{D_0 R_c}{D_0 + R_c} \tag{3.7}$$

In Eq. 3.7, A_p is the surface area of the particle; p_{ox} is the partial pressure of the oxidant species in the gas surrounding the combusting particle. Here the order with respect to the oxygen partial pressure is set as unity and this is the case with many computational models. There is evidence that the index is about 0.5–0.6 for most coal chars (Hurt and Calo 2001; Bews et al. 2001), and this may to be the case for biomass chars. Equation 3.6 takes into consideration the effects of both the kinetics and the diffusion.

For pf particles such as biomass or coal, the diffusion rate coefficient D_0 and the kinetic rate R_c is often calculated in commercial computer models (for example, the ANSYS FLUENT Manual 2014) as follows:

$$D_0 = C_1 \frac{\left[(T_p + T_\infty)/2 \right]^{0.75}}{d_p} \tag{3.8}$$

$$R_c = \eta \frac{d_p}{6} \rho_p A_g \left[A_i e^{-E_i/RT_p} \right] \tag{3.9}$$

where C_1 is an experimentally derived diffusion coefficient, T_∞ is the gas temperate and d_p is the particle diameter. The pre-exponential factor A_i and the activation energy E_i are for the intrinsic reactivity which can be obtained experimentally for a particular fuel. η is the effectiveness factor that represents the influence of the pore diffusivity and it can be expressed as a function of parameters such as the concentration of oxidant in the bulk gas, the bulk diffusivity, the porosity, tortuosity and the internal surface area of the char particle. The kinetic rate R_c is controlled by a combination of the intrinsic chemical reaction rate and the pore diffusion rate. Different regimes of char oxidation are identified. Regime I: chemical control occurs at relatively low temperatures with small particles such as in a TGA experiment where the effectiveness factor is unity, and the activation energy measured is the true value; Regime II is when there is both diffusion and chemical control and is found in many flames, the activation energy measured is half the true value, and Regime III occurs in hot flames where surface diffusion control takes place because the chemical reaction rate is very rapid.

While char reactivity for coals under pulverised fuel conditions are relatively well-established, only a very limited amount of data are available on biomass char burning rates or related data on the char surface area. Lang and Hurt (2003) presented data for relative reactivity measurements for a range of chars formed at 700–1,000 °C and found a correlation of reactivity and carbon content for pure chars, but that it was a very poor indicator of biomass reactivity. This was attributed to the catalytic behaviour of the biomass and subsequent studies have shown this is largely due to potassium. Prediction of the reactivity of a char in this temperature range is difficult because it is a function of many factors, composition, surface area and the potassium content. At higher temperature, in hot furnace flames, much of the potassium is lost by the char and the reactivity becomes close to that for coal chars and carbons, as discussed later. It is generally accepted that biomass chars are more reactive than coal char, particularly in the early stages of char conversion, because of its more disordered carbon structure and higher oxygen content. Unlike coal particles that soften and tend to become spherical, biomass particle tends to keep its original irregular shape during devolatilisation. Thus a larger surface area is often maintained at char combustion which results in a stronger oxygen flux into the biomass char particles and this contributes to a much higher char combustion rate.

From Eq. 3.5 it is seen than information is required on surface area through the lifetime of the particle, reactivity and the effect of annealing. This data is extremely difficult to obtain and there is very little information available for biomass chars. These surface areas are linked to the morphology of the chars produced.

It is well known that chars can be formed from the combustion of coal particles and also from sprays of heavy fuel oils. In the case of heavy oil the char usually consists of a hollow cenosphere consisting mainly of carbon and one cenosphere is produced per droplet; the mechanism being that outer surface of the droplet is carbonised and the inner contents evaporate via holes in the surface and are burned. Coal is similar in that the coal usually melts to give a liquid although the viscosity depends upon the coal type and the mineral content (which is higher than for oil). Eight types of coal chars have been identified with definitions being determined by the thickness of the wall and the internal porosity. Biomass chars are also different as illustrated in Fig. 3.7.

It seems that when a particle enters a flame the cellulose and hemicellulose are decomposed leaving a lignin skeleton as in Fig. 3.7a but on further heating the lignin can melt producing a char with a complex morphology. This will depend on the temperature it is exposed to and the amount and type of lignin in the biomass. A detailed study (Avila et al. 2011) has been made of chars produced from the pyrolysis of biomass particles of a number of different types in a packed bed at 1,000 °C. They found that biomass chars are significantly different from coal chars. That some chars retained their initial fibrous nature with thin walls (tenuisphere and tenui-network would be the coal char equivalent terms) in the case of Miscanthus, sunflower; thick walled spheres (crassisphere, crassi-network) were observed for Swedish wood and rapeseed; dense or solid particles were observed in only a few cases and these from smaller particles. The situation is different in fixed or fluid bed combustion where the particles would not be able to melt and the only form of char is a dense form like charcoal. Surface areas have also been measured at about 1,000 K by Murakami et al. (2013) for biomass chars.

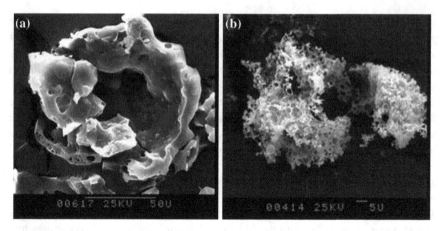

Fig. 3.7 Chars produced by wood **a** during burn-out, and **b** ash-rich char at latter stages of burn-out consisting largely of ash and about to fragment

Quantitative information on biomass char reactivity is available but not to the same extent as that for coal chars. The complicating factor with biomass chars is the significant influence of catalytic metals such as potassium (Wornat et al. 1995; Jones et al. 2007) which has not yet been completely resolved. Most of the data that have been obtained are for apparent chemical reaction rates largely using TGA measurements. Data obtained up to about 2007 have been compiled by di Blassi (2009) and this consisted of studies over the temperature range of 700–1,000 K which gave apparent activation energies of between 114 and 230 kJ/mol. Few measurements have been made at higher temperatures. The Sandia flame reactor operating at 1,500 K was used by Wornat et al. (1996) to determine the reaction rates of two biomass chars (pine and switchgrass) relative to coal chars, and concluded that the rates were 2–4 times faster. Drop tube furnace methods were used by Meesri and Moghtaderi (2003) and apparent reaction rate data was obtained at temperatures between 1,473 and 1,673 K for sawdust. They concluded the data obtained by both sets of experiments were consistent. Other combustion data is available from other sources such as iron ore sintering (Murakami et al. 2013) at about 1,000 K. Sorum et al. (2014) obtained the intrinsic reactivities for waste wood and newspaper obtaining values of about 1×10^{-6} g m^{-2} s^{-1} at 500 °C.

It is difficult to bring this data together to use to model biomass combustion systems. One unifying approach is to use the intrinsic reactivity although to do that means that the effects of surface area and catalysts have to be applied. Smith (1987) derived the intrinsic reactivity as a function of temperature for a range of pure carbons, namely $\rho_i = 3{,}050 \exp((-179.4)/RT)$kg/m^{2}s and this is shown in Fig. 3.8 as line 1. We have converted the data from di Blasi (2009) for a number of biomass fuels into the intrinsic reactivity form by assuming an average surface area of 100 m^2/g; but it should be noted that the surface varies over the combustion reaction and is different for each biomass. The values obtained are shown in Fig. 3.8 as region 2. Data for willow and Miscanthus from Jones et al. (2010) is shown marked as region 3. Both these sets

Fig. 3.8 Plot of intrinsic reactivities (kg/m² s) against 1/T (K). (*1*) Smith (1987) for a range of carbons. *Solid line* (*2*) data derived from Di Blasi (2009); (*3*) data for a number of biomass chars, Jones et al. (2010); (*4*) flame data (*filled circle*)

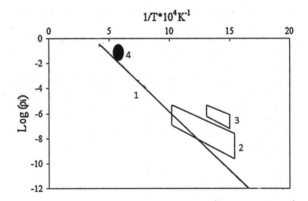

of data are for fuels containing substantial amounts of potassium (and other metals) which will catalyse the reactions and it is seen in the figure that these rates are faster than the uncatalysed reaction rates in Line 1. Washing reduces the reaction rate but not as far as to Line 1. Flame data from Wornat et al. (1996) and Meesri and Moghtader (2003) is shown as point 4. The latter derived an apparent activation energy of 89 ± 22 for Regime II combustion, which is equivalent to a true activation of 178 kJ/mol, which is very similar to the Smith value above.

It may be reasonable to assume that at lower combustion temperature the overall reaction consists of a carbon/oxygen reaction part and a metal (mainly potassium) catalyzed part, the latter dominating at this temperature. At higher temperatures, about 1,300 K from the data in Fig. 3.8 the carbon/oxygen reaction dominates and for many biomass chars the rate is about four times the Smith intrinsic rate. This would be suitable for pf combustion but difficulties remain for fluid and fixed bed combustion calculations where experimental data applicable to the case being considered would have to be used at least for the moment.

The combustion of larger particles in fluid and particularly fixed bed furnaces is largely controlled by heat transfer to the particle so a high level of accuracy for the oxidation kinetics is not necessary. But there are additional complications in that CO_2 and H_2O can react with the char produced although most of the reactions will take place in the gas phase. In large particles it is possible that there is an added impedance to the gas flow to the surface and this is from the formation of ash. This will very much depend on the system and will not be a factor in pf or fluidized bed combustion but may be a feature in systems in which there is little movement in the bed.

References

ANSYS FLUENT (2014) Ansys Inc, USA. www.ansys.com

Avila C, Pang CH, Wu T, Lester E (2011) Morphology and reactivity characteristics of char biomass particles. Bioresour Technol 102:5237–5243

Backreedy RI, Jones JM, Pourkashanian M, Williams A (2002) Modeling the reaction of oxygen with coal and biomass chars. Proc Combust Inst 29:415–422

Basu P (2013) Biomass gasification, pyrolysis and torrefaction: practical design and theory. Academic Press, Waltham

Bews IM, Hayhurst AN, Richardson SM, Taylor SG (2001) The order. Arrhenius parameters, and mechanism of the reaction between gaseous oxygen and solid carbon. Combust Flame 124:231–245

Blondeau J, Jeanmart H (2011) Biomass pyrolysis in pulverized-fuel boilers: derivation of apparent kinetic parameters for inclusion in CFD codes. Proc Combust Inst 33:1787–1794

Branca C, Di Blasi C (2013) A unified mechanism of the combustion reactions of lignellulosic fuels. Thermochemica Acta 565:58–64

Chemkin (2014) Reaction design. www.Reactiondsign.com

Chen Y, Charpenay S, Jensen A, Wojotowicz MA, Serio MA (1998) Modeling of biomass pyrolysis kinetics. Symp (Int) Combust 27:1327–1334

CRECK (2014) http://creckmodeling.chem.polimi.it

di Blasi C (2008) Modeling chemical and physical processes of wood and biomass pyrolysis. Energy Combust Sci 34:47–90

di Blasi C (2009) Combustion and gasification rates of lignocellulosic chars. Prog Energy Combust Sci 35:121–140

Faravelli T, Frassoldati A, Hemings EB, Ranzi E (2013) Multistep kinetic model of biomass pyrolysis in cleaner combustion. Springer, London, pp 111–139

Hayhurst AN (2013) The kinetics of the pyrolysis or devolatilisation of sewage sludge and other solid fuels. Combust Flame 160:138–144

Haynes BS, Newbury TG (2000) Oxyreactivity of carbon surface oxides. Proc Combust Inst 28:2197–2203

Hecht ES, Shaddix CR, Lighty JAS (2013) Analysis of the errors associated with typical pulverized coal char combustion modeling assumptions for oxy-fuel combustion. Combust Flame 160:1499–1509

Higman C, van der Burgt M (2008) Gasification, 2nd edn. Elsevier Inc, Amsterdam. ISBN: 978-0-7506-8528-3

Hurt RH, Calo JM (2001) Semi-global intrinsic kinetics for char combustion modelling. Combust Flame 125:1138–1149

Hurt RH, Sun JK, Lunden M (1998) A kinetic model of carbon burnout in pulverized coal combustion. Combust Flame 113:181–197

Jensen PA, Frandsen F, Dam-Johansen K, Sander B (2000) Experimental investigation of the transformation and release to gas phase of potassium and chlorine during straw pyrolysis. Energy Fuels 14:1280–1285

Jones JM, Patterson PM, Pourkashanian M, Williams A, Arenillas A, Rubiera F, Pis JJ (1999) Modelling NO_x formation in coal particle combustion at high temperature: an investigation of devolatilisation kinetic factors. Fuel 78:1171–1180

Jones JM, Darvell LI, Bridgeman TG, Pourkashanian M, Williams A (2007) An investigation of the thermal and catalytic behaviour of potassium in biomass combustion. Proc Combust Inst 31:1955–1963

Klass DL (1998) Biomass for renewable energy, fuels, and chemicals. Academic Press, San Diego

Kobayashi H, Howard JB, Sarofim AF (1977) Coal devolatilization at high temperatures. Proc Combust Inst 16:411–425

Lang T, Hurt RH (2003) Char combustion reactivities for a suite of diverse solid fuels and char-forming organic model compounds. Proc Combust Inst 29:423–431

Lasode OA, Balogun AO, McDonald AG (2014) Torrefaction of some Nigerian lignocellulosic resources and decomposition kinetics. J Anal Appl Pyrol. http://dx.doi.org/10.1016/j.jaap.2014.07.014

Lu H, Ip E, Scott J, Foster P, Vickers M, Baxter LL (2010) Effects of particle shape and size on devolatilization of biomass particle. Fuel 89:1156–1168

Ma L, Jones JM, Pourkashanian M, Williams A (2007) Modelling the combustion of pulverised biomass in an industrial furnace. Fuel 86:1959–1965

Meesri C, Moghtader IB (2003) Experimental and numerical analysis of sawdust-char combustion reactivity in a drop tube reactor. Combust Sci Technol 175:793–823

Murakami K, Sugawara K, Kawaguchi T (2013) Analysis of combustion rate of various carbon materials for iron ore sintering process. ISIJ Int 53(9):1580–1587

Niksa S (2000) Predicting the rapid devolatilization of diverse forms of biomass with bio FLASHCHAIN. Symp (Int) Combust 28:2727–2733

Peters B (2003) Thermal conversion of solid fuels. WIT Press, Southampton

Pickard S, Daood SS, Pourkashard M, Nimmo W (2013) Robust extension of the Coats-Redfern technique: reviewing rapid and reliable reactivity analysis of complex fields decomposing in inert and oxidizing thermogravimetric analysis atmospheres

Piriou B, Vaitilingom G, Veyssière B, Cuq B, Rouau X (2013) Potential direct use of solid biomass in internal combustion engines. Prog Energy Combust Sci 39:169–188

Ranzi E, Frassoldati A, Stagni A, Pelucchi M, Cuoci A, Faravelli T (2014) Reduced kinetic schemes of complex reaction systems: fossil and biomass-derived transportation fuels. Int J Chem Kinet 46:512–542

Saddawi A, Jones JM, Williams A, Wojtowicz MA (2010) Kinetics of the thermal decomposition of biomass. Energy Fuels 24:1274–1282

Saddawi A, Jones JM, Williams A (2012) Influence of alkali metals on the kinetics of the thermal decomposition of biomass. Fuel Process Technol 104:189–197

Sheng CD, Azevedo JLT (2002) Modelling biomass devolatilization using the chemical percolation devolatilization model for the main components. Proc Combust Inst 29:407–414

Smith IW (1987) The intrinsic reactivity of carbons to oxygen. Fuel 57:409–414

Sørum L, Campbell PA, Erland N, Haugen L, Mitchell RE (2014) An experimental study of the reactivity of cellulosic-based chars from wastes. Fuel 130:306–314

van Lith SC, Jensen PA, Frandsen FJ, Glarborg P (2008) Release to the gas phase of inorganic elements during wood combustion. Part 2: influence of fuel composition. Energy Fuels 22:1598–1609

Weber R (2008) Extracting mathematical exact kinetic parameters from experimental data on combustion and pyrolysis of solid fuels. J Energy Inst 81:226–233

Williams A, Jones JM, Ma L, Pourkashanian M (2012) Pollutants from the combustion of biomass. Prog Energy Combust Sci 38:113–137

Wornat MJ, Hurt RH, Yang NYC, Headley TJ (1995) Structural and compositional transformations of biomass chars during combustion. Combust Flame 100:131–143

Wornat MJ, Hurt RH, Davis KA, Yang NYC (1996) Single-particle combustion of two biomass chars. Symp (Int) Combust 26:3075–3083

Chapter 4
Pollutant Formation and Health Effects

Abstract This chapter deals with pollutant formation and the consequential health effects. The major features of pollutants arising from biomass combustion are discussed. These include smoke, Unburned Hydrocarbons (UBH), volatiles, Polycyclic Aromatic Hydrocarbons (PAH), Nitrogen Oxides (NO_x), other nitrogenous pollutants, sulphur, chlorine compounds and dioxins. Trace metals and particularly the interaction of K–Cl–S chemistry and aerosol emissions are discussed.

Keywords Pollutant formation · Smoke · NO_x · PAH · Trace metals

4.1 General Feature of Pollutants Arising from Biomass Combustion

Combustion of biomass results in a number of toxic products such as carbon monoxide, incompletely combusted organic compounds and smoke which are formed from the C, H, O content of the biomass. There are also acidic species formed from the N, S and Cl functional groups as well as metallic species from the trace elements. The routes leading to the formation of these are shown in Fig. 4.1, which parallels the main combustion paths shown in Fig. 3.1. The nature and extent of the pollutants produced depends on the biomass use, the combustor design and control measures used. Generally large combustion units with carefully monitored combustion control form low levels of pollutants whereas small units with poor mixing, short residence times and little or no control systems tends to form high levels of pollutants.

© The Author(s) 2014
J.M. Jones et al., *Pollutants Generated by the Combustion of Solid Biomass Fuels*,
SpringerBriefs in Applied Sciences and Technology,
DOI 10.1007/978-1-4471-6437-1_4

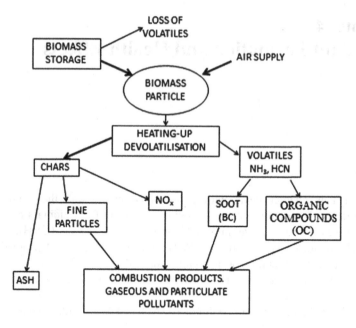

Fig. 4.1 Diagram of the pollutant formed during biomass combustion

4.2 Smoke, UBH, Volatiles, PAH and Odour

Soot formed by the combustion of biomass is essentially the same as that formed from fossil fuels with only slight differences in composition. The general term 'smoke' includes all particulate material emitted such as ash, inorganic aerosols, PAH, tar particles, small particles of fragmented char and soot. Soot consists of agglomerated chains of carbonaceous spherules of elemental carbon (EC). These basic carbonaceous particles formed together with some fine particles of inorganic compounds have condensed organic compounds (OC) associated with them. PAH is distributed between the Black Carbon (BC) and OC fractions of the soot depending on its molecular weight which affects their volatility. The nature of the smoke emitted depends on the fuel being burned, the conditions under which it is formed and surrounding atmosphere before being emitted into the atmosphere. These processes involved are condensation, coagulation and agglomeration.

There have long been major concerns about the health effects of smoke (Naeher et al. 2007; Bollinger et al. 2009) and more recently about the effects on the climate (Crutzen and Andreae 1990; Bond et al. 2013). Much of the developed world has legislation controlling air quality such as the UK Clean Air Act resulting from smog deaths in the 1950s. The effects on health are now most severe in Developing Countries where very large numbers of small, low combustion residence-time cookers and heaters are still used. Around 3 billion people cook and heat their homes using open fires and simple stoves burning biomass (wood, animal dung and crop waste) and coal. There has been concern about the effects of wild fires, as the frequency of these is predicted to increase with climate change resulting in a further increase in global temperature. Smoke emissions are correlated with millions of deaths which result every year (WHO 2014). About 4 million people die prematurely from illness attributable to the household air pollution from cooking with solid fuels. More than 50 % of premature deaths among children younger than 5 years old are due to pneumonia caused by particulate matter (soot) inhaled from household air pollution.

Smoke from biomass combustion has recently assumed greater importance from a climate perspective. Whilst considerable attention has been devoted to the effect of the principal Greenhouse Gases including carbon dioxide, methane and nitrous oxide, it has recently been demonstrated that the role of both black carbon (BC) and organic carbon (OC) are substantial (Bond et al. 2013). BC consists of the whole light absorbing fraction of the carbonaceous aerosols which mainly consists of EC. The ratio of BC to the co-emitted OC is of interest in terms of the effect of soot particles on climate change. Particles with a higher ratio of BC absorb heat rather than reflecting it, affect cloud albedo, and affecting the reflectivity of snow and ice. Some studies suggest that biomass particles can exhibit higher ratios of BC to OC in comparison with conventional fuels. About 47 % of the global emissions of BC is from the combustion of biomass.

The nature of BC and OC in the soot have been analysed by various methods. By definition, light reflectance methods are typically used for BC, however techniques such as TGA give a value for the fixed elemental carbon which approximates well to the BC. Analysis of the OC can be achieved by techniques such as pyrolysis GC-MS or solvent extraction of the extracted organic compounds followed by GCMS analysis.

The health effects and climate effects from the OC component are related. The results are complicated by the fact that the organic material consists of the precursors to the formation of soot and by incompletely combusted fuel and emissions are accompanied by NO_x, SO_x and inorganic aerosols that can have an influence on cloud formation. Climate forcing from co-emitted species were estimated and used in the model by Bond et al. (2013). When the effects of short-lived co-emissions, including cooling agents such as sulphur oxides, are included in the net forcing, both fossil and biomass fuels have a positive climate forcing during the first year after emission, i.e. climate warming. For older technology diesel engines and possibly residential biofuels, warming is strong enough that eliminating all short-lived emissions from these sources would reduce net climate forcing, that is, produce cooling. When wild fire and open burning emissions are included, the best estimate of net industrial-era climate forcing by all short-lived species from black-carbon-rich sources becomes slightly negative. The relative effects of these factors are shown in Fig. 4.2.

Fundamental studies have been made of the formation of soot from the combustion of simple hydrocarbons in order to develop soot forming mechanisms (Frenklach and Wang 1991; Colket and Hall 1994; D'Anna and Violi 2005; Ranzi 2009). Extensive studies have been made of the properties of the soot particles

Fig. 4.2 Climate 'Forcing' for different biofuel combustion activities. For each activity the *upper bar* shows the positive radiative forcing due to black carbon, the *second bar* shows the negative radiative forcing due to negative effects due to co-emitted species, and the *third bar* shows the overall effect (information from Bond et al. 2013)

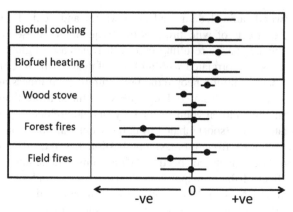

Climate forcing per activity per mass of fuel burned

produced and a comprehensive coverage is given in the Proceedings of a Workshop held in 2007 (Bockhorn et al. 2009). These mechanistic studies have included aliphatic, unsaturated and aromatic compounds including coal and most recently from solid biomass combustion. Models have been proposed for the formation of soot from wood combustion (Fitzpatrick et al. 2008) whereby pyrolysis products from the lignin components are considered to be responsible for much of the soot formation. An issue still under investigation is whether this influences the BO/OC ratio.

The extent and nature of the soot produced from the combustion of biomass is a function of the burner design and thus the combustion conditions and the post-flame temperature-time history. Many studies have been undertaken particularly of fixed bed grate combustors with small or medium thermal capacity, and also from moving bed combustors and suspension or pf combustion. A considerable body of information exists on the related process of smoke from the combustion of simple hydrocarbon fuels, but whilst there are some similarities to biomass there are significant fundamental differences.

The particulate from biomass consists of conventional smoke, oxygenated aromatic compounds and pyrolytically derived PAH, residual unburned char particles and products that are condensed on the particulates. The nature of the resultant aerosols depends on the way in which they are formed and on their reactant history as they emerge from the combustion zone and are emitted into the atmosphere and indeed on the subsequent atmospheric chemistry. The products obtained are also a function of the composition of the biomass, the moisture content and the combustion stoichiometry.

The mechanism proposed by Fitzpatrick et al. (2008) is set out in Fig. 4.3 and some details related to it in Fig. 4.4. In Fig. 4.3 three routes are indicated.

- Route 1 leads to the formation of small molecules such as CO and small hydrocarbon molecules largely from the pyrolysis of hemicellulose and cellulose. These then undergo complete combustion to produce CO_2 or reform to produce larger molecules such as benzene and hence to PAH via the HACA mechanism.

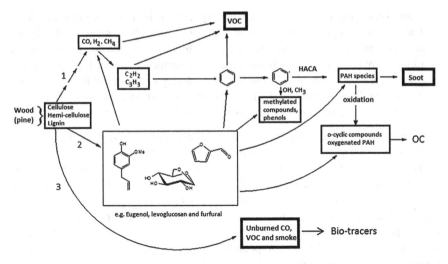

Fig. 4.3 Pathways leading to smoke and unburned hydrocarbon formation (based on Fitzpatrick et al. 2008)

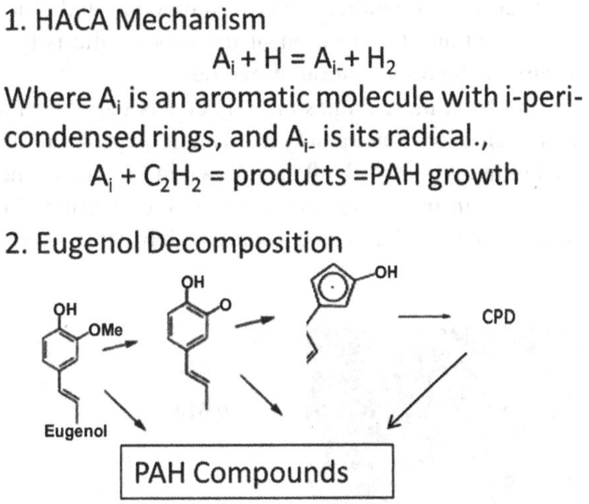

Fig. 4.4 Outline of the HACA mechanism (*top*), and the initial steps in the decomposition of a model lignin decomposition product, eugenol (*bottom*)

1. HACA Mechanism
$$A_i + H = A_{i-} + H_2$$
Where A_i is an aromatic molecule with i-peri-condensed rings, and A_{i-} is its radical.,
$$A_i + C_2H_2 = products = PAH\ growth$$

2. Eugenol Decomposition

PAH Compounds

This is an acronym for 'H-abstraction-C_2H_2-addition' and is a repetitive reaction outlined in Fig. 4.4.

- Reaction Route 2 results directly in the formation of larger molecules from cellulose including furfural, or eugenol and related compounds from lignin. These react further as indicated in Fig. 4.3.
- Route 3 is a path by which some of the intermediate species are not completely combusted and escape from the combustion unit into the atmosphere. These are characteristic of biomass combustion and are termed biotracers or sometimes

biomarkers. One of the most abundant species in biomass combustion, from wood or herbaceous materials, is levoglucosan as well as some of the lignin pyrolysis products. Another characteristic biotracer is retene. Retene is an alkylated phenanthrene and is believed to be formed by from the resin in the wood. It is usually formed during the smouldering phase of combustion. These should not occur under perfect combustion condition and only arise from imperfect mixing in simple cook stoves or wild land biomass fires.

The details of the HACA mechanism are set out in Fig. 4.4. Also shown are main reaction steps for a typical lignin pyrolysis product, eugenol.

The mechanism depends on the combustion conditions and the post-flame temperature-time history. During the combustion of the biomass both pyrolysis and oxidative pyrolysis of the biomass components occurs with the release of tars and vapours. The initial products obtained are a function of the composition of the biomass. There are significant differences in biomass pyrolysis products and their tendency to form soot and PAH compounds. This in turn influences the tendency to form BC/OC described earlier.

Figure 4.4 shows a typical sooty flame of a wood pyrolysis product-furfural and Fig. 4.5b shows a computed opposed diffusion flame calculated using Chemkin (2014) and the CRECK (2014) combustion mechanism. The formation of benzene, naphthalene and anthracene is seen together with fluorene. Other PAH compounds are formed and the position of the soot profile is based on the assumption that it occurs by the dimerisation of pyrene.

Soot from these flames initially consists spherical particles of about 2 nm in the soot nucleation zone rapidly growing to particles of about 50 nm in the diameter in the luminous part of the flame shown in Fig. 4.5a. These particles will form chains higher up in the flame depending on the 'sootiness' of the flame, i.e. the number density of the spherical particles which determines their rate of agglomeration.

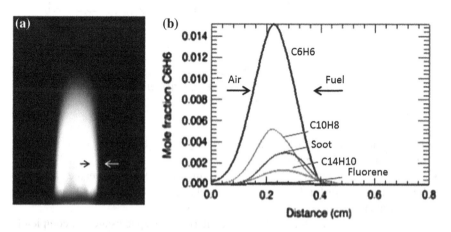

Fig. 4.5 **a** Diffusion flame of furfural burning on a wick burner. **b** Opposed diffusion flame of furfural computed using the Chemkin (2014) Opp-Dif Program and CRECK (2014) model

Fig. 4.6 Electron microscope image of soot formed from biomass combustion

This may also be determined by the aromatic nature of the soot particles and the number of reactive centres present. Figure 4.6 shows a typical electron microscope image of soot formed by such a flame. There is virtually no visual difference in the nature of soot formed by biomass combustion and from the model simple fuels such as ethene.

Polycyclic aromatic compounds (PAC) and polycyclic aromatic hydrocarbons (PAH) are of special interest in biomass combustion as they comprise the largest group of environmental carcinogens, or if not carcinogenic are synergists; they are important as soot precursors as previously described. PAC have been shown to be products of pyrolysis of a range of natural products which make up biomass, including carbohydrates, lignin, lipids, plant resins and terpenes.

Polycyclic aromatic hydrocarbons (PAH) are the most examined group among the PAC; they consist of conjoined benzene rings, either linearly annelated as in anthracene, angularly annelated as in phenanthrene, or peri-condensed as in pyrene. Five-membered ring structures such as fluoranthene are also possible. Examples of PAH structures are anthracene, phenanthrene, pyrene and fluoranthene.

PAC relate to PAH through the substitution into one or more benzene rings with sulphur-, nitrogen- or, particularly in the case of biomass, oxygen to produce heterocyclic structural analogues or isosteres. Examples are dibenzothiophen, carbazole and dibenzofuran. These considerations along with the multiplicity of possible alkylated derivatives result in the number of isomers for even a relatively low molecular weight to be very large and the PAC or PAH mixtures from biomass combustion are extremely complex. To simplify comparisons between PAH

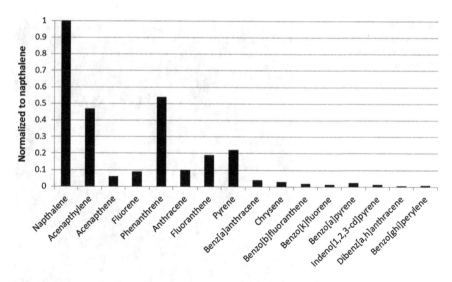

Fig. 4.7 Illustration of the type of PAH compounds formed by the combustion of wood and by coal

sources the US Agency for Toxic Substances and Disease Registry proposed 16 priority PAH to represent the very many possible PAH.

Figure 4.7 shows the types of the PAH compounds formed during wood combustion. Their concentrations are determined by their thermodynamic stability which vary according to ring size and structure.

The EPA Priority PAH compounds with their molecular weights in parentheses are:

1. naphthalene (128),
2. acenaphthylene (152),
3. acenaphthene (154),
4. fluorene (166),
5. phenanthrene (178),
6. anthracene (178),
7. fluoranthene (202),
8. pyrene (202),
9. benz[a]anthracene (228),
10. chrysene (228),
11. benzo[b]fluoranthene (252),
12. benzo[k]fluoranthene (252),
13. benzo[a]pyrene (252),
14. indeno[1,2,3-cd]pyrene (276),
15. dibenz[a,h]anthracene (278),
16. benzo[ghi]perylene (276).

Other compounds that are observed are the methylnaphthalene and methylphenanthrene compounds together with other isomers. In addition to PAH adsorbed on soot particles and in the gas phase, condensed PAH can also be emitted as particulates from biomass combustion but especially domestic appliances (Eriksson et al. 2014).

The complexity of PAC mixtures emitted during biomass combustion necessitates analytical separation methods of the highest resolution. Capillary column gas chromatography (GC) on columns coated with SE-52 (5 % phenyl methylsiloxane) stationary phase of fractions separated from the extracted PAH mixture by liquid chromatography (LC) has proved an effective procedure; an alternative method of deriving the PAC fraction is by rapid heating and injection into the GC column (pyrolysis-GC). In either case the most effective GC detector is the mass spectrometer since the mass spectra of PAC are well understood as consisting mainly of an intense molecular ion, and for alkylated PAH an $(M - 15) +$ ion. However, such spectra do not allow PAH isomers to be distinguished and identifications by MS must be supplemented by GC retention data. The PAH elution order is in sequence of volatility and retention indices relative to naphthalene, phenanthrene, chrysene and picene standards for over 400 PAC on SE-52 have been published.

Biological activity varies between isomers and minor constituents, e.g. 5-methylchrysene and dibenzo[def,p]chrysene are potent carcinogens whereas naphthalene is toxic but not currently considered carcinogenic. The biological activity is now studied by the use of bacterial mutation of PAH rather than animal testing. The mechanism of carcinogenic action has been shown to be via oxidation by liver enzymes to yield dihydrodiol epoxides. These compounds then form DNA adducts which result in cell mutation and eventual malignancy.

4.3 Nitrogen Oxides (NO_x) and Other Nitrogenous Pollutants

Nitrogen oxides (NO and NO_2) have been identified as an important precursor to acid rain and to ozone formation. There are also significant health effects from these. Nitrogen dioxide forms nitric acid with water which can irritate the lungs and lower resistance to respiratory infections including influenza and other respiratory illnesses. The primary pollutant formed by combustion NO and it is converted to NO_2 by reaction with O_2. The rate of conversion depends on the temperature and their concentrations so the NO_x emitted from most combustion systems consists mainly of NO but this is slowly converted in the atmosphere to NO_2. The Threshold Limit Value (TLV) for NO is 25 ppm and that for NO_2 is 3 ppm so clearly good ventilation and dispersion of the combustion products is important.

There are three main reaction routes leading to the formation of NO_x:

(i) thermal NO_x from high temperature oxidation of atmospheric N_2 (the Zeldovich mechanism),
(ii) prompt NO_x from the reaction of fuel-derived radicals with atmospheric N_2, and
(iii) fuel-NO_x from oxidation of nitrogen which is chemically bound in the fuel.

Fig. 4.8 NO formation pathways from both thermal and fuel-N routes

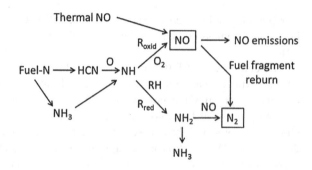

N_2O can also be formed in lean flames or by the reduction of NO by carbon present in the char. Nitric oxide once formed will combine with oxygen to form NO_2 so the major nitrogen pollutants produced is NO_x (NO + NO_2) and although the amount of NO_2 discharged is usually small, about 5 %, all the NO initially formed is converted in the atmosphere to entirely to NO_2. Thus the NO_x emission is reported in terms of NO_2 equivalent, and the N_2O is not included in this total. Depending on the combustion conditions it is also possible that 8NH_3, HCN and nitroso-compounds are present in smaller amounts in the flue gases. The main pathways to NO_x formation are shown in Fig. 4.8.

Thermal NO_x is formed typically at reaction zone temperatures of about 1,800 K. Thus with advanced multistage low NO_x burners the formation of the thermal NO_x can be small. Prompt NO_x is produced largely in rich flames in the presence of hydrocarbons which usually contribute to a minor portion of the total NO_x formation. When firing 100 % biomass, the flame temperatures are relatively low and thus both thermal and prompt NO_x may be small. Therefore, the fuel NO_x, which is produced by the oxidation of nitrogen contained in the fuel, often becomes the main contributor to the overall NO_x formation. Thus control fuel-nitrogen input to the system, i.e. choice of fuel, is an effective way to reduce the overall NO_x emissions from a power plant. The nitrogen content of biomass is usually very small and depends on the source. Trees contain about 0.1–0.5 wt% N although tropical trees contain more fuel-N. Bark and leaves contain up to a few % N. Proteins are the main source of –N in biomass. But the compounds change during the seasons and on senescence convert to cyclic –N compounds. Ideally biomass crops for combustion should be carbon rich and with low N and low trace elements.

The favoured sources of biomass for combustion are currently SRC (short rotation coppice) willow and the perennial grass, Miscanthus. Forestry co-products are another important fuel source and other perennial grasses such as switchgrass. These species have the advantage that they are lignin (carbon) rich (SRC and woody materials) or have minimal N content due to nutrient recycling to the underground rhizome (e.g. in the case of perennial grasses such as Miscanthus).

For perennial crops, material for combustion is usually harvested following extensive senescence and drying. The senescence processes result in chlorophyll breakdown, protein breakdown, and a variable amount of remobilisation to non-harvested fractions. Factors controlling some of these processes are not fully understood, however final composition (of N-containing compounds) will influence the form of residue and type of emissions i.e. HCN from cyclic N, and NH$_3$ and HCN from protein and amide compounds, and subsequent NO and N$_2$O formation in both cases (Bai et al. 2013). For example, the N-content in Miscanthus at harvest in February/March is lower than in June/July. What is not known is how the different N species (inorganic nitrate and ammonium ion, amino compounds, heterocyclic purines, pyrimidines and pyrroles vary with time. But potentially the choice of harvest time can be used to minimise NO$_x$ emissions.

Firing biomass instead of coal in a coal fired furnace can reduce the overall NO$_x$ emission from the plants although the reduction on a thermal basis is not substantial. The important issue in such converted plants is whether the reduction in NO emission is sufficient to be able to dispense with the need for additional NO$_x$ reduction plant. However the reduction in S-content would alleviate the need for SO$_x$ removal processes.

There is variability both in total N content and in composition between biomass types, for example, natural senescence leads to a lower N content due to reduced protein-N, although leaving a relatively higher proportion of heterocyclic aromatic compounds. Nitrogen oxide emissions upon combustion can be controlled by informed choice of biomass material and/or appropriate pre-treatments with a target of minimising heterocyclic N compounds.

In Fig. 4.8 it is shown that the initial pyrolysis step produces both HCN and NH$_3$ which ultimately react to give NO, N$_2$O and N$_2$. Poor mixing in the reaction zone results in both these intermediate compounds can be emitted in small quantities in the flue gases. In order to minimise the amount of NO produced it is necessary to understand the chemistry of the first step to HCN and NH$_3$ and the factors controlling their subsequent reactions so that N$_2$ is produced rather than NO. During devolatilisation the pyridinic and pyrrolic nitrogen is released as HCN, while proteins and amino acids can also release NH$_3$. The structure of some typical amino acids found in biomass and a generic protein are shown in Fig. 4.9.

Recently a number of studies of the pyrolysis of model nitrogen compounds have been made and a picture is being built up of the way in which they react and can form HCN and NH$_3$. Generally the cyclic compounds decompose by losing the side chain which then forms NH$_3$ whilst the cyclic unit produced HCN. In the case of proteins both HCN and NH$_3$ are produced but the ratio depends on the temperature.

The way in which nitrogen partitions between volatiles and char is important particularly as this can impact on the way in which low NO$_x$ burners function. In the first stage where the volatiles are produced under fuel-rich conditions much of the fuel-nitrogen is converted (approximately 90 %) to N$_2$. The char is burned in the second stage is burned under lean conditions is largely converted to NO. Thus if the nitrogen is release as volatiles the staged low NO$_x$ burner is more effective. Many models assume that the C/N molar ratio remains the same in the volatiles and the char as it is in the original biomass. The experimental evidence suggests that this may be the case for woods and Miscanthus but is not the case for Olive

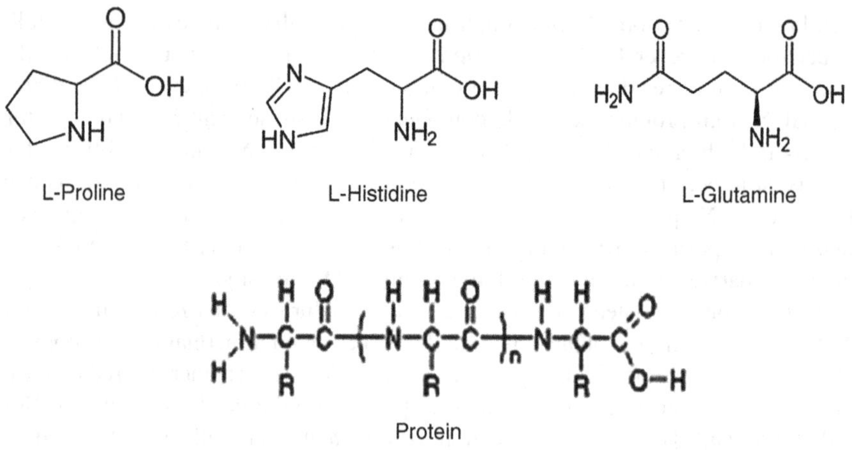

L-Proline   L-Histidine   L-Glutamine

Protein

Fig. 4.9 Typical organic nitrogen compounds in biomass

waste and PKE where a greater proportion is lost in the volatiles (Stubenberger et al. 2008; Di Nola et al. 2010; Darvell et al. 2012).

In the subsequent reactions of the HCN and NH_3 intermediates, De Soete (1975) derived equations to give the rate of the oxidation to NO (R_{oxid}) and reduction (R_{red}) to N_2. For HCN these are respectively:

$$R_{oxid} = 1 \times 10^{10} \, X_{HCN} \, [X_{O2}]^n \exp\{-280.4/RT\}s^{-1} \qquad (R\,4.1)$$

$$R_{red} = 3 \times 10^{12} \, X_{HCN} \, X_{NO} \exp\{-251.2/RT\} \qquad (R\,4.2)$$

and for ammonia

$$R_{oxid} = 4 \times 10^6 \, X_{NH3} \, [X_{O2}]^n \exp\{-133.9/RT\} \qquad (R\,4.3)$$

$$R_{red} = 1.8 \times 10^8 \, X_{NH3} \, X_{NO} \exp\{-113.0/RT\} \qquad (R\,4.4)$$

where R_i are the reaction rates, X_i the mol fraction, and n is the reaction order, which is zero for X_{O2} greater than 3 mol% oxygen. The relative rate of formation of NO from the two routes is very temperature-dependent and is subject to experimental error. Another option is to use a full chemical kinetic model. A useful simplified model has been given by Hansen and Glarborg (2010).

Nitrous oxide (N_2O) is also produced by the combustion of biomass as previously stated. The formation of NO and N_2O from char is complex and an important case is from large scale fluidised bed combustion. This is a significant issue because N_2O has 289 times the Global Warming Potential (GWP) of carbon dioxide over a 20 year period (UN IPCC 4th Assessment Report 2007). The main source of N_2O is agriculture responsible for 78 % in 2011 in the UK and which includes the production of biomass fuels. Combustion of all types of fuel was responsible for 11 % in 2011. It is necessary to verify that replacing fossil fuels with bio-energy crops will lead to a net reduction in Greenhouse Gas (GHG) emissions. N_2O from growing energy crops is highly dependent on the amount of nitrogen fertiliser used. It has been stated that the use of 1st generation biofuels that require N fertiliser could cause as much global warming as that avoided by use of the biofuel in place of fossil fuels (Smith et al. 2012).

N_2O is produced in flames is due to (Kramlich et al. 1989):

(a) Oxidation of amine radicals

$$NH + NO = N_2O + H \qquad (4.5)$$

where the amine radical NH is formed by the decomposition of ammonia and hydrogen cyanide produced from fuel-nitrogen.
(b) In fluidized bed furnaces where combustion takes place in a fluidised bed containing quantities of char with temperatures of approximately 850 °C. Under these conditions the char partially reduces NO produced to N_2O.

$$C \text{ (solid)} + 2\,NO = N_2O + CO \qquad (4.6)$$

N_2O can also be produced in very lean conditions by the reaction

$$O + N_2 + M = N_2O + M \qquad (4.7)$$

Fuel composition can have a significant effect on N_2O emission. It has been seen that co-firing biomass with coal in a fluidised bed results in lower N_2O emissions just as SO_2 is reduced. The higher oxygen content of biomass fuels relative to fossil fuels enhances production of isocyanic acid (HNCO), which is a precursor for N_2O along with HCN. However the overall picture with regard to N_2O formation is not yet clear but it is believed that when waste-derived biomass is used the effects of preservatives and glues sometimes present can result in increased N_2O production.

4.4 Sulphur, Chlorine Compounds and Dioxins

Biomass tends to contain only small amounts of sulphur and chlorine. When biomass is substituted for coal in co-firing the lower sulphur content is advantageous in respect of SO_2 control. The concentration of chlorine in biomass ranges from 0.2 to 2 %, however the higher concentrations found in straws is a major problem.

Both Cl and S are present both as organo-compounds and as inorganic salts, mainly the latter particularly as potassium salts in solution in the xylem and phloem. Both of these are released partly during devolatilisation with the more refractory salts being retained and released during char combustion by sublimation or evaporation of the oxides, hydroxides or chlorides. Cl which is released as both

KCl and HCl depending on the temperature gives an acidic and corrosive vapour which reacts with other metals forming gas phase species and then aerosols during the cooling processes. This also leads to deposits on the furnace walls and causes corrosion.

Apart from the aerosols, dioxins and furans can be formed from the HCl present reacting with aromatic species such as phenols (Lavric et al. 2004). The term 'dioxins and furan' is the abbreviated title for a group of chlorinated aromatic compounds of which 2,3,7,8-tetrachloro-p-dibenzo-dioxin (2,3,7,8,TCDD) is the most toxic and these compounds can be present in the gaseous emissions and the ash from the furnaces. They can have cancerous and non-cancerous effects for which the emission limits have decreased over the last decade because of issues relating to emissions from some industrial processes and incinerators burning high chlorine-containing wastes. The formation of dioxins is complex, and in the early days of co-firing there was concern that the addition of a high chlorine-containing biomass such as straw to the coal would result in dioxin formation. This could result because the biomass decomposition products contain phenolic species which in the hotter regions of flue gases can form phenoxy radicals which can combine to give dioxin precursors. Chlorination catalysed by metals in the fly ash can give polychlorinated dioxins. However as long as combustion is complete, so decreasing the amount of C/H/O species in the flue gases and the ash, then the dioxin release is also minimised and this environmental problem is minimal.

4.5 Metals, K–Cl–S Chemistry and Aerosol Emissions

Biomass contains a wide range of trace metals, some adventitious metals taken in from the soil which can become an issue in contaminated land. The most important metal is potassium which varies from several thousand ppm in energy crops to several weight % (dry basis) in certain agricultural residues. Other important species are sodium, calcium and silica. On combustion these metals are released forming inorganic particles which are then exhausted into the atmosphere as pollutants or deposited in the furnace walls causing slagging, fouling and corrosion. This process is illustrated in Fig. 4.10.

Particulate emissions consist of both carbonaceous products and ultrafine inorganic components. The quantity of each group depends upon the nature of the biomass and the combustion conditions. Usually the ultrafine ($<d = 100$ nm) components near the immediate vicinity of the flame zone mainly consists of inorganic species, more refractory species such as silica tend to remain in the ash particle and are released later. As far as emissions are concerned the various metal species are released and interact in the gas phase where there is sufficient reaction time available for them to come to equilibrium, the products then depending on the temperature. The most important species present would be K (and KO and KOH), Na and Ca compounds as well as the volatile metals Pb and Zn. The other trace metals such as Cr, Mn, Fe and the non-metallic species such as Si, S and P also

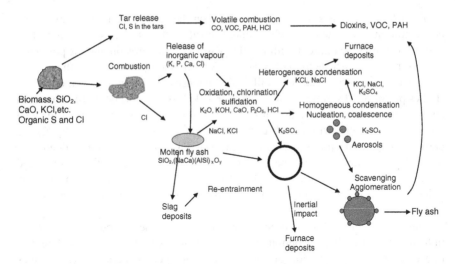

Fig. 4.10 Diagrammatic presentation of metals release (adapted from Bryers 1996)

play a role. Their composition would depend on the temperature and on reaching cooler parts of the combustion chamber condense and are emitted as aerosols. The material deposited in the furnace causes slagging, fouling and corrosion; this aspect is a major topic considered in many reviews.

The nucleation of K_2SO_4 and the condensation of KCl also significantly contribute to the increased deposition. The release of the potassium that exists in biomass particles is usually in the form of KCl gas (the amount determined by the content of Cl in the biomass) or KOH. Experimental investigations show that the release of KCl during devolatilisation may occur at two temperature stages. At a lower temperature of about 300–500 °C, a small amount of KCl is released from the biomass particles whilst a more significant amount is released above 700 °C.

Aerosols generated from biomass combustion can be a result of an incomplete combustion such as soot, PAH, fragments of ash (Eriksson et al. 2014). The majority of the fine aerosols are produced as a result of reaction of alkaline metals and chlorine present in the biomass which leads to the formation of sticky and corrosive components and submicron particulates, such as KCl and K_2SO_4. The form in which the metal species are released from combustion of biomass depends on the relative contents of the potassium and chloride in the biomass. Once released as a part of the volatiles the potassium and chlorine undergo a series of chemical and physical transformation steps to form more stable KCl and K_2SO_4 compounds in gaseous phase and then through nucleation and condensation to form aerosols. The detailed kinetics of the sulphation of alkali spices involves hundreds of intermediate reactions and reduced reaction mechanisms are usually employed (Glarborg and Marshall 2005). Fine particles can penetrate the lungs more deeply than larger particles and have significant health effects. There have been a number of publications concerned with the modelling process and these are discussed in Chap. 7.

The KCl in the gas stream may condense on the K_2SO_4 nuclei forming a high concentration of fine aerosols which may deposit on the surface of the combustion system or it may be emitted with the flue gases causing erosion and air pollution. At present, the transformation of SO_2 to S (VI) and the mechanisms of the nucleation of the sulphuric acids are still under intensive investigation. Experimental studies performed by Jiménez and Ballester (2008) on the formation of submicron aerosols from biomass combustion shows that in a SO_2 rich environment the KCl particles, which otherwise are observed in the flue gases, were replaced by K_2SO_4 particles. In order to minimise the effect of slagging, coal ash is often added with the fuel as a 'getter' where in particular the calcium aluminium silicates react and absorb the potassium (Wu et al. 2013). However this ash can also form fine particles because of fragmentation and attrition and should be subject to a particle removal system.

Thus the chemical composition of the biomass and the combustion conditions influence the overall distribution of pollutants formed. At ignition and refuelling conditions the formation of soot and organic species predominate, whereas under efficient flaming conditions the ash particles remain in the ultrafine state. In smouldering conditions soot and ash, but mainly the latter are formed and influence the agglomeration properties depending on the amount of channelling that takes place (Torvela et al. 2014).

References

Bai J, Yu C, Li L, Wu P, Luo Z, Ni M (2013) Experimental study on the NO and N_2O formation characteristics during biomass combustionl. Energy Fuels 27:515–522

Bockhorn H, D'Anna A, Sarofim AF, Wang H (eds) (2009) Combustion generated fine carbonaceous particles. In: Proceedings of an international workshop, Anacapri 2007, KIT Scientific Publishing, Karlsruhe

Bolling AK, Pagels J, Yttri KE, Barregard L, Sallsten G, Schwarze PE, Boman C (2009) Health effects of residential wood smoke particles: the importance of combustion conditions and physicochemical particle properties. Part Fibre Toxicol 6:29

Bond TC, Doherty SJ, Fahey DW, Forster P, Berntsen T et al (2013) Bounding the role of black carbon in the climate system, a scientific assessment. J Geophys Res Atmos 118:5380–5552

Bryers RW (1996) Fireside slagging, fouling, and high temperature corrosion of heat transfer surfaces due to impurities in steam-rising fuels. Prog Energy Combust Sci 22:29–120

Colket MB, Hall RJ (1994) In: Bockhorn H (ed) Soot formation in combustion: mechanisms and models. Springer, Berlin, pp 442–470

Crutzen PJ, Andreae MO (1990) Biomass burning in the tropics: impact on atmospheric chemistry and biogeochemical cycles. Science 250:1669–1678

D'Anna A, Violi A (2005) Detailed modeling of the molecular growth process in aromatic and aliphatic premixed flames. Energy Fuels 19:79–86

Darvell LI, Brindley C, Baxter XC, Jones JM, Williams A (2012) Nitrogen in biomass char and its fate during combustion—a model compound approach. Energy Fuels 26:6482–6491

De Soete GG (1975) Overall reaction rates of NO and N_2 formation from fuel nitrogen. Proc Combust Inst 15:1093–1102

Di Nola G, de Jong W, Spliethoff H (2010) TG-FTIR characterization of coal and biomass single fuels and blends under slow heating rate conditions: Partitioning of the fuel-bound nitrogen. Fuel Process Technol 91:103–115

Eriksson AC, Nordin EZ, Nyström R, Pettersson E, Swietlicki F, Bergvall C, Westerholm R, Boman C, Pagels JH (2014) Particulate PAH emissions from residential biomass combustion: time-resolved analysis with aerosol mass spectrometry. Environ Sci Technol 48:7143–7150

Fitzpatrick EM, Jones JM, Pourkashanian M, Ross AB, Williams A, Bartle KD (2008) Mechanistic aspects of soot formation from the combustion of pine wood. Energy Fuels 22:3771–3778

Frenklach M, Wang H (1991) Detailed modeling of soot particle nucleation and growth. Proc Combust Inst 23:1559–1566

Glarborg P, Marshall P (2005) Mechanism and modeling of the formation of gaseous alkali sulfates. Combust Flame 141:22–39

Hansen S, Glarborg P (2010) A simplified model for volatile-N oxidation. Energy Fuels 24:2883–2890

Hansson K-M, Amand L-K, Habermann A, Winter F (2003a) Pyrolysis of poly-L-leucine under combustion-like conditions. Fuel 82:653–660

Hansson K-M, Samuelsson J, Amand L-E, Tullin C (2003b) The temperature's influence on the selectivity between HNCO and HCN from pyrolysis of 2,5-diketopiperazine and 2-pyridone. Fuel 82:2163–2172

Jimenez S, Ballester PM (2008) Vaporization of trace elements and their emission with submicrometer aerosols in biomass combustion. Energy Fuels 22:2270–2277

Kramlich JC, Cole JA, McCarthy JM, Lanier WS, McSorley JA (1989) Mechanisms of nitrous oxide formation in coal flames. Combust Flame 77:375–384

Lavric ED, Konnov AA, De Ruyck J (2004) Dioxin levels in wood combustion—a review. Biomass Bioenergy 26:115–145

Naeher LP, Brauer M, Lipsett M, Zelikoff JT, Simpson CD, Koenig JQ, Smith KR (2007) Woodsmoke health effects: a review. Inhalation Toxicol 19:67–106

Ranzi E (2009) Detailed kinetics of real fuel combustion: main paths to benzene and PAH formation. In: Bockhorn H, D'Anna A, Sarofim AF, Wang H (eds) Combustion generated fine carbonaceous particle. KIY Scientific Publishing, Kahrlstuhr, pp 99–124

Smith KA, Mosier AR, Crutzen PJ, Winiwarter W (2012) The role of N_2O derived from crop-based biofuels, and from agriculture in general, in Earth's climate. Philos Trans R Soc B 367:1169–1174

Stubenberger G, Scharler R, Zahirovic S, Obernberger I (2008) Experimental investigation of nitrogen species release from different solid biomass fuels as a basis for release models. Fuel 87:793–806

Torvela T, Tissari J, Sippula O, Kaivosoja T, Leskinen J, Virén A, Lähde A, Jokiniemi J (2014) Effect of wood combustion conditions on the morphology of freshly emitted fine particles. Atmos Environ 87:65–76

UN IPCC (2007) 4th assessment report, United Nations

WHO (2014) World Health Organisation. Geneva, Switzerland

Wu H, Bashir MS, Jensen PA, Sander B, Glarborg P (2013) Impact of coal ash addition on ash transformation and deposition in a full scale suspension-firing boiler. Fuel 113:632–643

Chapter 5
Emissions from Different Types of Combustors and Their Control

Abstract Emissions from different types of combustors and their control methods are outlined. These include emissions from fixed and travelling bed combustors and emissions from large industrial combustion plants. Emissions from wild Fires are also considered.

Keywords Emission factors · Bed combustors · Pulverised fuel combustion

5.1 Emissions from Biomass Combustion

An important aspect of the formation of pollutants from the combustion of biomass is an accurate prediction of the actual amounts of the pollutants formed from a particular type of combustor using a specific type of biomass. Unfortunately at present there is no exact way in which this can be done because of the wide range of equipment used and variability in the operating procedures. Combustion equipment of all types may be subject to legislation for emission of toxic gases (NO_x, SO_x, CO) and particulates, initially for PM_{10} but now for finer particulates ($PM_{2.5}$) and perhaps even finer particles (PM_1) in future. An understanding of the emissions and the way in which they can be controlled is of vital importance in terms of local and national pollution. All the pollutants discussed in the previous chapter will be formed, but the amount of each pollutant is determined by the nature of the fuel, any pre-treatment that it has received and the mode of combustion and in particular the supply of the combustion air, any pollution control methods and the means of exhaust and dispersion. A number of pollution control methods can be applied such as cyclone units which is the easiest to use but in some cases additional abatement technology may be required such as ceramic filters, electrostatic precipitators and bag filters. For each pollutant an emission factor (EF) can be determined experimentally and these emission factors are important in estimating national and global emission inventories.

J.M. Jones et al., *Pollutants Generated by the Combustion of Solid Biomass Fuels*,
SpringerBriefs in Applied Sciences and Technology,
DOI 10.1007/978-1-4471-6437-1_5

5.2 Emissions from Fixed and Travelling Bed Combustors

Combustion in fixed beds can be undertaken in a number of ways and this determines the amount of pollutants produced. The use of fuels such as irregular sized wood logs or briquettes made of sawdust or straw as well as the influence of factors such as bed size and the level of moisture being important. Fuel reloading which causes intermittent combustion is a major issue. Biomass can be burned in fixed bed combustors with thermal outputs of a few kWth to units of 1 MWth or larger especially if the unit combines suspension firing with grate firing at the bottom of the combustion chamber. Larger units will use a travelling grate system or equivalent in sizes up to several MWth. On a global basis the majority of biomass is burned in small cookers or heaters. These small combustion units have, because of their size, limited residence time for the combustion process and for mixing of the fuel and air. The combustion sequence in these units starts with ignition of the fire (or start-up), followed by flaming combustion, smouldering as the original charge of fuel burns out and then more fuel has to be added or refueled and then the process is repeated. The ignition process is a major source of smoke since the combustion chamber is not hot enough to ensure burnout of the released volatiles, when the combustion chamber has heated up during the flaming stage combustion is more efficient. This is also the case for the smouldering phase. But when a batch of cold fuel is added during refuelling the combustion chamber is temporarily cooled with the emission of smoke and volatiles and so on as the cycle is repeated. Thus these combustors are a major source of smoke. These types of units are widely used in developing countries with large populations such as Africa, India and China are often used indoors. Usually this involves the improvement of mixing or reduction in heat loss which increases the combustion temperature and increases the NO_x as a consequence.

Estimation of emissions from all types of bed combustion is difficult because, whilst the gaseous combustion occurring in the furnace chamber above the surface of the bed is relatively easy to simulate, the bed itself is not easy to simulate. The bed consists of irregular shaped particles where the initial fuel size may be tens or hundreds of millimeters or more. Whilst the term 'fixed' is used, combustion is not uniformly continuous and the particles are constantly moving. The fuel is often fed on a batch basis, the bed is not fixed in space and holes can appear through the bed in a random manner, known as 'channelling'. Thus emissions can only be given on an average basis. The introduction of pellet fuels in the last decade has simplified these calculations since the fuel is more uniform and combustion much more consistent. Thus a number of research groups have been able to carry out simulation studies of biomass combustion for pellet combustion as discussed in Chap. 6.

Combustion in the bed is only the beginning of the overall combustion process since the bed-generated products can undergo reaction in the upper part of the combustion unit above the bed as shown in Fig. 2.2a. This is effectively a secondary combustion chamber where additional (secondary) air is added. This reaction zone can be used to burn out particulate soot, PAH and other organic material.

However nitric oxide formation may be increased because of the oxygen added and careful design such as staging can minimise this process (Fitzpatrick et al. 2009; Liu et al. 2010). Basic combustion units of the type shown in Fig. 2.1 are not able to do this and the only methods of emission control are by choice of the fuel and by increasing the efficiency. As far as the fuel is concerned then the use of charcoal is best because the volatiles content in the fuel is very much lower and hence smoke and OC emissions are reduced. But much depends on the way the charcoal is produced: the choice of feedstock and the degree of reactivity imparted by the method of production are important. The use of torrefied fuel is also an option since this is a partially devolatilised biomass with properties intermediate between biomass and charcoal. The efficiency of the stoves can be improved by the use of insulation and the method of heat transfer to the cooking utensil. Improving the performance reduces the amount of emission per unit of material cooked.

A number of studies have been made of emission factors for different types of combustors using a variety of fuels. Some of these are given in Table 5.1. But it must be noted that these are average values over the whole cycle, the emission factor being highest at the start of combustion or during refuelling. Many test methods use the emission values for only the flaming combustion stage and consequently underestimate the total emission during the combustion cycle.

The quantity of dioxins (PCCD) is within usually acceptable limits and is about 0.3 μg/kg. The amount of SO_x depends on the soil in which it is grown but typically the EF is about 0.03 g/kg. The EF for NO_x similar to that for oil fired burners and it is much higher than for gas fired units. In principle the emission factors can be converted into the concentration of the pollutants in the flue gases but in reality there are wide variations in the amount of excess air added and it may change throughout the burning cycle. Some typical flue gas concentrations are given in Table 5.2. These concentrations are in terms of dried flue gas compositions.

Table 5.1 Typical emission factors (EF) for wood combustion

Type of unit	CO (g/kg)	NO_x (g/kg)	PM_{10} (g/kg)	CH_4 (g/kg)	Non-methane VOC (NMVOC) (g/kg)	PAH (mg/kg)
Open cook stove	100–350	8	20	10	30	20
Modern closed stove 8 kWth	50–130	10	5	1	10–40	1
Automatic pellet boiler 25 kWth	10–15	3–12	0.5–1.5	0.2	1	0.1
Moving grate 250 kWth	2.5–8	1.8–3.0	1.0	0.001	0.01	0.1

Sources included Johansson et al. (2004), Nussbaumer et al. (2008), Roden et al. (2009), Zhang et al. (2010), Orasche et al. (2012), Elsasser et al. (2013)

Table 5.2 Typical emission levels from the combustion of biomass for various sources

Type of unit	Village cook stove	Residential boiler, 2–10 kW	Pellet stoves, fixed grate 2–25 kWth	Fixed grate[a] 20 kW–2.5 MWth	Travelling grate[a] 150 kW–15 MWth	Fluidized bed[a] 100 MWth	Co-firing: coal/biomass[a] 1 GWth
Fuel	Wood	Wood chips, or Miscanthus	Wood pellets	Most biomass	Straw	Most biomass	Most biomass
NO_x mg/Nm³ – depends on fuel-N content	100–500	100–500	100–500	150–400	150	150	250
Particulate (mg/Nm³)	5,000	20–100	20	1	1	5	15
VOC (mg/Nm³)	100	10	10	10	10	15	0.1
CO (mg/Nm³)	250–6,000	250–800	200–800	130–600	150	5	100
Reference O_2 level (mol%)	11	6	11	11	11	11	6

Ratings are given in thermal capacity. Average values are given for the whole combustion cycle
[a] Assumes pollution control equipment is fitted

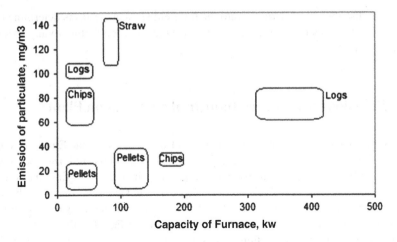

Fig. 5.1 Emission of total particulate (mg/m^3) plotted against thermal capacity of the biomass furnace firing wood pellets, wood chips, logs and straw (based on Ehrlich et al. 2007a, b)

The level of the particulates emitted largely depends on the type of fuel being burned and the form it is in and this is illustrated in Fig. 5.1.

The chemical and physical properties of the fuel are important as well as the design and thermal capacity of the appliance. For the smaller appliances it is seen that wood pellets give lower emissions. This is consistent with the operation of fixed bed combustion (Fig. 2.2a) where it is obvious that large pieces of fuel in a small combustor permit the volatiles to escape without being completely combusted and so pellets are best.

The health hazards are well documented in relation to indoor heating by biomass, which is a method used by 3 billion people. More recently attention has been directed to the emission from domestic and small commercial units used in developed countries. Legislation is being introduced to reduce NO$_x$ and smoke generally. In particular attention has been directed to the emission of PM$_{2.5}$ and even finer particles using detailed analytical method to study the properties during the different phase of combustion-ignition, flaming combustion, smouldering and then refuelling. Examples are Elsasser et al. (2013) and Torvela et al. (2014). This latter work showed that the formation of particles of ash and soot particles are largely separate processes. They showed that the particles consisted of inorganic species and carbonaceous matter. It was found that zinc oxide particles with an average diameter of <13 nm acted as seeds for the condensation of inorganic vapours and organic material, forming ash particles with a nested structure. The outer layer was composed mainly of alkali salts. Two modes of particle size distribution were observed. The main particle type found in the ultrafine particle size mode (<100 nm) was ash. Decreasing the combustion air increased the formation of soot and condensable organics into the particulate matter which was found mainly in the accumulation particle size mode (>100 nm). Thus the experimentally observed distribution of particles into the different size categories is determined by a number of factors making it difficult to attribute emission factors to any specific furnace or fuel.

5.3 Emissions from Large Industrial Combustion Plant

Emissions from large plants are usually controlled by National Emission regulations. These plants include travelling and fixed grate units, fluidized bed combustors, pulverised fuel or suspension firing which includes co-firing and 100 % biomass combustion. The levels of emissions permitted are determined by the size of the plant under legislation according to the country or region such as the EU. Emission factors are determined in many cases by the fact that these plants have pollution control equipment fitted so as not to exceed their permitted emission level. Some pollutants are not covered in this way and the emissions are determined by the type of fuel and plant operation. Pollution control measures, low NOX_x burners and particulate removal, have been described by Speight (2013). Table 5.3 gives, for example, the EU emission limits for biomass fired plants.

For example, when cofiring was introduced Spliethoff and Hein (1998) showed that the effect of co-combustion with biomass has a beneficial effect on emissions in pulverized fuel furnaces. Since biomass in most cases contains considerably less sulphur than coal, an increasing biomass share in the thermal input makes the SO_2 emissions decrease proportionally. SO_2 can also partly be captured in the ash by the alkaline–earth fractions of the biomass ash. Due to the high volatile content of the biomass, low NO emissions can be achieved both by air staging and by over-fire air injection (OFA) above the burner zone. The amount of unburned carbon (UBC) which is a waste product from pf combustion varies depending on the size of the initial fuel particles and the amount of moisture present—large/wet particles are incompletely combusted and result in increased UBC. Since it most economic to use large wet particles of wood, it is desirable to be able to determine the optimum size. This applies not only to pf combustion but to all biomass applications. Fluidized bed combustion is increasingly being for biomass combustion because it offers the potential of burning high fuel-N containing material which may have a high moisture level, but it is not without problems (Khan et al. 2009). In general but particularly in relation to fluidized bed combustion it is noted that the amount of moisture has to be controlled to prevent excessive combustion temperatures causing slagging.

Table 5.3 EU emission limits for biomass fired plants, at 6 % oxygen. (*Units* mg/m^3)

Solid fuels (gas)	Current plants	Future plants
NO_x	200	150
SO_2	200	150
Total particulate	100	20
CO	Not set	Not set

5.4 Wild Fires

The emissions from forest, bush and grass fires and from waste agricultural burning and land clearing is considerable as shown in Fig. 4.2. It contributes about 2,500 TgC per annum (about 2.5 Gtoe). It varies greatly on a year-by-year basis but appears to be on the increase as predicted by climate change models. This, together with biomass burning for cooking and heating, is a major contributor of particles in the atmosphere as well as high levels of organic matter, are included in the total and with consequential effects (see Sect. 4.2).

Studies have shown that most of the particles from biomass burning were found to be of a size less than 1 μm in diameter. Other pollutants such as CO and NO_x are also formed and can be very important in the vicinity of the fires (Jaffe et al. 2008).

The residues remaining after incomplete combustion of vegetation (char) can contribute significantly to the carbon content of soils. Char-C is considered biologically inert relative to other forms of organic C in soils; however, the degree of biological inertness is likely to be a function of the completeness of the combustion process. Char may contribute significantly, up to 30 wt% to soil organic carbon (SOC) because of the widespread occurrence of wildfires in many parts of the world.

References

Ehrlich C, Noll G, Kalkoff WD (2007a) Determining PM-emission fractions (PM_{10}, $PM_{2.5}$, $PM_{1.0}$) from small-scale combustion units and domestic stoves using different types of fuels including bio fuels like wood pellets and energy grain. DustConfd 2007. How to improve air quality. In: International conference, Maastricht, The Netherlands, April 2007. www.dustconf.com

Ehrlich C, Noll G, Kalkoff W-D, Baumbach G, Dreiseidler A (2007b) PM_{10}, $PM_{2.5}$ and $PM_{1.0}$-emissions from industrial plants—results from measurement programmes in Germany. Atmos Environ 41:6236–6254

Elsasser M, Busch C, Orasche J, Schon C, Hartmann H, Schnelle-Kreis J, Zimmermann R (2013) Dynamic changes of the aerosol composition and concentration during different burning phases of wood combustion. Energy Fuels 27:4959–4968

Fitzpatrick EM, Bartle KD, Kubachi MI, Jones JM, Pourkashanian M, Ross AB, Williams A, Kubica K (2009) The mechanism of the formation of soot and other pollutants during the co-firing of coal and pine wood in a fixed-bed combustor. Fuel 88:2409–2417

Jaffe DA, Hafner W, Hand D, Westerling WL, Spracklen D (2008) Interannual variations in $PM_{2.5}$ due to wildfires in the Western United States. Environ Sci Technol 42:2812–2818

Johansson LS, Leckner B, Gustavsson L, Cooper D, Tullin C, Potter A (2004) Emission characteristics of modern and old-type residential boilers fired with wood logs and wood pellets. Atmos Environ 38:4183–4195

Khan AA, de Jong W, Jansens PJ, Spliethoff H (2009) Biomass combustion in fluidizes bed boilers: potential problems and remedies. Fuel Process Technol 90:21–50

Liu H, Qiu G, Shao Y, Riffat SB (2010) Experimental investigation on flue gas emissions of a domestic biomass boiler under normal and idle combustion conditions. Int J Low-Carbon Technol 5:88–95

Nussbaumer T, Czasch C, Klippel N, Johansson L Tullin C (2008) Particulate emissions from biomass combustion in IEA countries. In: Survey on measurements and emission factors. IEA Bioenergy Task 32 Zurich. ISBN 3-908705-18-5

Orasche J, Seidel T, Hartmann H, Schnelle-Kreis J, Chow JC, Ruppert H, Zimmermann R (2012) Factors and characteristics from different small-scale residential heating appliances considering particulate matter and polycyclic aromatic hydrocarbon (PAH)-related toxicological potential of particle-bound organic species. Energy Fuels 26:6695–6704

Roden CA, Bond TC, Conway S, Pinel ABO, MacCarty N, Still D (2009) Laboratory and field investigations of particulate and carbon monoxide emissions from traditional and improved cookstoves. Atmos Environ 43:1170–1181

Speight JG (2013) Coal-fired power generation handbook. Scrivener, MA

Spliethoff H, Hein KRG (1998) Effect of Co-combustion of biomass on emissions in pulverized fuel furnaces. Fuel Process Technol 54:189–205

Torvela T, Tissari J, Sippula O, Kaivosoja T, Leskinen J, Virén A, Lähde A, Jokiniemi J (2014) Effect of wood combustion conditions on the morphology of freshly emitted fine particles. Atmos Environ 87:65–76

Zhang X, Chen Q, Bradford R, Shariffi V, Swithenbank J (2010) Experimental investigation and mathematical modelling of wood combustion in a moving grate boiler. Fuel Process Technol 91:1491–1499

Chapter 6
Mathematical Modelling

Abstract Modelling of biomass combustion using Computational Fluid Dynamics (CFD) is covered in the chapter. The Reynolds-Averaged Navier-Stokes equations are outlined as well as turbulence-chemistry interactions. Modelling pulverised biomass particle combustion is outlined including the sub-models. These are particle motion, heat transfer, devolatilisation and char combustion. Modelling pulverised fuel co-firing in power stations is next considered using the methods outlined previously. Modelling fixed and fluidised bed combustion is then described. The application of these methods to model the emission of nitrogen and sulphur oxide emissions and aerosol pollutants is outlined.

Keywords Combustion modelling techniques · Pollutant formation · Metal aerosols

6.1 Modelling Biomass Combustion Using Computational Fluid Dynamics

Computational Fluid Dynamics (CFD) modelling has been developed as an important tool for the design, optimization and prediction of pollutants from large pulverised coal combustion plants and these techniques have been extended to co-firing coal and biomass (Williams et al. 2001). Some work has been devoted to modelling firing 100 % pulverised biomass systems but little is published because they are not widely used at present. CFD studies of emissions from fixed and fluidised bed combustion of biomass have been published over the last decade as well as some work on incineration where the fuel is assumed to have a biomass-like composition. Over the past decade, combustion modelling using CFD has been transformed from qualitative modelling to quantitative predictions in many engineering applications. A number of commercial CFD software packages are now available that are capable of providing reasonable simulations for solid fuel combustion, with user defined specific sub-models which are employed in the coal/biomass combustion modelling applications. Modelling techniques have been developed for various degrees of complexity. Currently, three-dimensional simulations for full-scale furnaces with millions of computational cells are achievable.

© The Author(s) 2014
J.M. Jones et al., *Pollutants Generated by the Combustion of Solid Biomass Fuels*,
SpringerBriefs in Applied Sciences and Technology,
DOI 10.1007/978-1-4471-6437-1_6

CFD combustion modelling is based on the solution of a set of fundamental conservation equations that govern the general reacting flows incorporating fuel specific processes. The bases of these are set out in a number of text books as well as the solution techniques available (For example: Andderson et al. 2012; Chung 2002; Wendt 2009). Solid fuel combustion involves the follow five main tasks:

1. Gas phase flow and volatile combustion modeling, including pollutant predictions,
2. Particle fluid dynamics modelling and inter-phase heat/mass transfer,
3. Devolatilisation modelling,
4. Heterogeneous char combustion modelling, and
5. Reactions of inorganic species including condensation and deposition.

CFD modelling was initially used for geometrically simple gaseous and spray flames with single step combustion chemistry. These spray models were developed into models of pulverised coal combustion because of the similarity of the combustion process as well as the handling of the particulate systems. Application to fixed bed combustion was only developed recently and initially took the approach that a simple fixed bed model produced combustible gases above the bed that then reacted in the gas phase. These models have become more sophisticated including the dynamic motion of the bed. These models have been extended to fluidised bed combustion but little work has been undertaken on the combustion of biomass.

We have a good knowledge about modelling pulverised coal combustion and it is generally accepted that many aspects of the combustion of pulverised biomass are similar to that of coal. Thus these models were adapted and/or with some modifications to the biomass combustion. Here two fundamental modeling approaches exist. These are the Eularian-Eularian and the Eularian-Largragian approaches. In the Eularian method, the discrete solid phase is also considered as a continuum. Therefore the Eulerian conservation equations are solved for both the gas and the suspended particulate phases and the model can include all the stages of a particle combustion process, particle drag and turbulent dispersions, etc. The distinctive benefit of the Eulerian-Eulerian approach is its computational efficiency which is the result from the elimination of the particle tracking and iterations and corrections between dispersed particles and the continuum phase. The Eularian-Lagrangian method, which is the predominant simulation procedure for pulverised solid fuel combustions, solves the motion of combusting particles by tracking them separately in a Lagrangian reference frame, whilst the gas phase is solved in the Eulerian domain. Most solid fuel combustion submodels are designed for this approach.

In the Eularian-Lagrangian approach, the fluid flow and volatile combustion are modelled by solving the time-averaged equations of global mass, momentum, energy, and species mass fractions within an Eulerian reference frame. Turbulence models and turbulence-chemistry interaction schemes are incorporated to simulate the effect of turbulence on the reacting flow. These equations may be solved by using general CFD numerical solution technique such as the control volume method or the finite element method. Pulverized fuel particles are typically treated as dilute and dispersed particulate flows and they are modelled in a Lagrangian reference frame where the particle motion and reaction are tracked step-by-step by time throughout the combustion system. The assumption behind this approach is that the distance between any two fuel particles is large that any chance of particle interactions is small and therefore may be ignored. With this assumption, the movement and the heterogeneous reaction of each combusting particle stream is tracked individually by a set of Lagrangian formulations. The coupling of the gaseous and the solid phases is established through the source/sink terms of mass, momentum and energy in the Eulerian gas-phase conservation equations, and the overall solution involves a multi-loop iterative procedure.

A general solution strategy for modelling the co-firing of biomass and coal within the Eularian-Lagrangian framework is illustrated as in Fig. 6.1. The overall iterative solution process begins by specifying the appropriate properties of the fuel and oxidant used in the combustion system and the boundary conditions, such as the flow rate of both the air and the fuel. Then the governing partial differential equations for the reacting gas phase flow may be solved according to the given boundary conditions. The particle specific submodels are used to calculate the particle trajectories, heating up, drying, volatile release and char reactions. These calculations generate and provide updates to the source terms, as well as other relevant parameters and physical properties that are required in the governing partial differential equations for the gas flows. The convergence of the iteration is checked; if the residue of the solution is large than a prescribed tolerance then the calculation should continue with the updated source terms and physical properties until the convergence criteria is

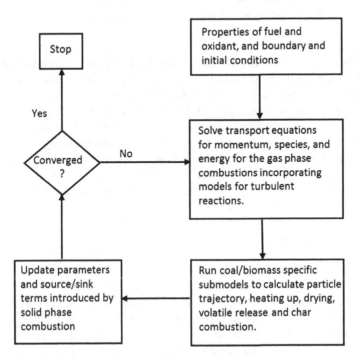

Fig. 6.1 Eularian-Lagrangian approach for modelling co-firing of biomass and coal

satisfied. Most general purpose CFD software packages have these numerical facilities built in. However, those sub-models that are unique to the physical and chemical processes of the solid fuel combustion, in particular for the biomass fuels which are usually quite different from that of coal, have to be specially developed.

6.1.1 Reynolds-Averaged Navier-Stokes Equations

The Reynolds-Averaged Navier-Stokes equations are commonly employed for turbulent reacting flows. Steady state Reynolds-Averaged Navier-Stokes equations and the continuity equation may be written in a Cartesian co-ordinate system as follows:

$$\frac{\partial \rho u^i u^j}{\partial x^j} = -\frac{\partial p}{\partial x^i} + \frac{\partial}{\partial x^j}\left[\mu\left(\frac{\partial u^i}{\partial x^j} + \frac{\partial u^j}{\partial x^i}\right)\right] + \frac{\partial \tau^{ij}}{\partial x^j} + \rho g^i + F^i \qquad (6.1)$$

$$\frac{\partial \rho u^i}{\partial x^i} = S_m \qquad (6.2)$$

where u^i, p, ρ and μ are the mean fluid velocity components, pressure, density and molecular viscosity of the fluid flow, respectively, g^i is the gravitational acceleration and F^i is the source term for the external force, such as the force arisen from the interaction with the dispersed biomass (or coal) particles in the flow. Further,

$$\tau^{ij} = -\rho \overline{u'^i u'^j} \qquad (6.3)$$

is the Reynolds stress tensor. This results from the Reynolds average of the Navier-Stokes equations and it represents the effects of the turbulence in the fluid flow, and the overbar denotes a Reynolds average and the dash represents the fluctuating part of the velocity of the fluid flow. This is where a turbulence model comes into the governing fluid flow equations that will be discussed later, which links the Reynolds stress tensor τ^{ij} to the global mean fluid velocity u^i.

In addition to the Navier-Stokes equations, which prescribe the motion of the fluid flow, the energy equation is used to govern the heat transfer within the gas phase and between the gas and the solid phases. The energy equation can usually be written in the following form:

$$\frac{\partial u^i (\rho e + p)}{\partial x^i} = \frac{\partial}{\partial x^i}\left[k_{eff}\frac{\partial T}{\partial x^i} + D_s + \Phi\right] + S_e \qquad (6.4)$$

where e, T and k_{eff} are the internal energy, temperature and the effective heat conductivity of the fluid flow, respectively; D_s represents the energy transfer due to species diffusions and Φ is a diffusion term; and S_e is the energy source term such as heat from a chemical reaction. In coal/biomass combustion simulations, this should also include the heat transfer between the gas and solid particles that are responsible for the heating and cooling of coal or biomass particles.

In a chemical reacting system, the changes in the concentrations of each of the species in the system should follow the basic conservation law and this may expressed in the form of transport equations as follows:

$$\frac{\partial\left(\rho u^i Y^k\right)}{\partial x^i} = -\frac{\partial J^k}{\partial x^i} + S_k, \quad k = 1, 2, \ldots, n-1 \qquad (6.5)$$

where Y^k, J^k and S_k denote the mass fraction, the diffusion flux and the production and destruction of the species k, respectively, during the combustion processes, and n denotes the total number of chemical species involved in the combustion system and this is usually dependent on the knowledge of the composition of the volatiles and the reaction mechanism employed in the numerical simulation. For coal/biomass combustion, S_k should include the source/sink terms resulting from the drying, devolatilisation and char combustion of the coal and biomass.

Equations 6.1–6.5 form the basic set of governing equations for the gaseous phase combustion modelling of the coal and biomass combustion. Additional char combustion models are required providing source term for heat and species mass transfer. Further, appropriate turbulence models have to be used to calculate the Reynolds stress tensor.

It should be noted that there are three fundamental numerical techniques that may be employed to simulate turbulent reacting flows with various degrees of detail along with different computational costs. They are the Direct Numerical Simulation (DNS) technique, the Large Eddy Simulation (LES) technique and the Reynolds-Averaged Navier Stokes (RANS) technique, each of which employs different forms of the fundamental governing partial differential equations.

The DNS technique employs the full Navier-Stokes equations without using any models for the turbulence. This technique requires the computational cells to be small enough to resolve the full spectrum of the turbulent eddy that exist in the fluid flow. This usually results in the use of such a large number of computational cells that makes it not practical for most engineering modelling applications. The LES directly solves for the large scale turbulence that is resolved by the computational grid employed and the remaining sub-grid turbulent eddies are modeled using the so-called subgrid turbulence models. The idea of this is that small scale eddies are mainly responsible for the turbulence energy dissipation and they can be modelled relatively independent to the boundary conditions. Therefore, the size of the computational grid employed in the LES can be much larger than that used in DNS. This can substantially reduce the number of grids required in a CFD model.

The RANS employs the Reynolds-Averaged Navier-Stokes equations given in this section where all the spectrum of the turbulent eddies in the system, from large to small scales, are modelled with appropriate turbulence models. For engineering applications, a number of different turbulence models are now available, each involving different levels of complexity but inevitably all have their limitations in use. With the one equation mixing-length turbulence model being the simplest and the Reynolds Stress Model (RSM) the most complex, the k-ε turbulence models are the most commonly used in various engineering applications, including combustion simulations.

The k-ε model is based on the Boussinesq eddy-viscosity hypothesis which relates the Reynolds stress tensor to the mean fluid velocity through the following equations:

$$\tau^{ij} = -\frac{2}{3}\rho k \delta_j^i + \mu_t \left(\frac{\partial u^i}{\partial x^j} + \frac{\partial u^j}{\partial x^i} \right) \tag{6.6}$$

$$\mu_t = c_\mu \rho \frac{k^2}{\varepsilon} \tag{6.7}$$

where μ_t is known as the turbulence viscosity and c_μ is an empirically derived constant and usually takes the value of 0.09. The turbulent kinetic energy k and the dissipation rate ε in the k-ε model are further calculated by two more model transport equations.

As we can see from Eqs. 6.6 and 6.7, the turbulence viscosity μ_t in the k-ε model is a scalar which assumes an isotropy in the turbulence. It is because of this assumption that makes the k-ε model frequently over predict the turbulence viscosity in the situations where non-isotropic turbulence prevails. In pf combustion applications, this occurs where there is a strong swirling flow, such as in the burner near-field and in T-fired furnaces. It should be noted that due to the apparent failings of the standard k-ε model in terms of its inability to account for strong rotational strain or shear, various nonlinear modifications have been developed. Nevertheless, the standard k-ε model is still very widely used in various combustion models and this is largely because of the relative robustness in computations.

For strong swirl flows, a more elaborate RNG k-ε model may be preferred which is also easy to implement and the convergence is robust. The RNG k-ε model is very similar in its formulations to the standard k-ε model. However, an extra term is introduced in the transport equation for the dissipation rate to account for the effect of the anisotropic large-scale eddies giving substantial improvement in the predictions of swirling flows, and flows with a large degree of strain and/or a large boundary curvature.

The Reynolds-Stress Models (RSMs) have emerged as strong candidates for use in industrial turbulent flow calculations. Since the RSMs are second-order closure models which model the different evolutions of the components of the Reynolds stress, $\tau^{i,j}$, separately with different transport equations. Therefore, they are potentially superior to the simpler eddy-viscosity models, such as the k-ε models. However, it is worth mentioning that various Reynolds Stress Models, including those of the simplified Algebraic Stress Models comprise a group of complicated formulations that require a potentially excessive amount of computational effort for the simulation of three-dimensional fluid flows. In addition, it is reported that not all the fluid flow simulations can benefit from the RSMs to justify the high computational costs. This is why even though more sophisticated models have been proposed, the k-ε model with its various modifications is still the most popular turbulence model in use.

6.1.2 Turbulence-Chemistry Interactions

In laminar combustion flows, chemical reactions are usually governed by the Arrhenius equation that can be expressed in general form as:

$$k = AT^{\alpha}e^{-E/RT} \tag{6.8}$$

where k is the rate constant, α is the non-dimensional temperature exponent, A and E are the pre-exponential factor and the activation energy of the reaction, respectively, and R is the universal gas constant. Once the reaction path for the combustion has been specified then the species production/reduction, as well as heat generations, can be calculated according to the local concentrations of the reactants and the reaction rate with Eq. 6.8. A number of kinetic compilations are available such as Chemkin (Reaction Dynamics 2014) and the CRECK (2014) data.

Pulverised solid fuel combustion is usually in a highly turbulent environment and turbulent gaseous species mixing may play a significant, if not a dominant, role in turbulent combustions of coal and biomass. Chemical reactions in non-premixed combustion, such as the volatile combustion, occur only when the fuel and oxidant come into contact molecularly. Therefore, the relative importance of turbulence on the rate of combustion is determined by the speed at which chemical species come into molecular contact, i.e. the speed of mixing, and the speed at which chemical reaction takes place in the combustion. Volatile combustion is fast and the time scale of the chemical reaction is much smaller than that of the turbulent mixing. Therefore the reaction is mainly mixing controlled. In such cases, the complex chemical kinetic rates can usually be neglected and the fast reaction assumptions, such as 'equilibrium', 'mixing is burnt' or one or two step global reaction may be employed. The two most frequently used fast reaction modeling approaches are the eddy-breakup model based, such as the eddy-dissipation model, and the other is the mixture fraction model based.

The eddy-dissipation model is a type of eddy-breakup model which calculates the turbulent chemical reaction rate in association with the eddy dissipation/mixing time scale, k/ε. In implementing this model, three different reaction rates may be calculated for each species, namely, the Arrhenius reaction rate, the rate due to the dissipation of turbulent reactant eddies and the rate due to the dissipation of turbulent production eddies. The net rate of production of the species due to reaction is given by the smallest of the three. The eddy-dissipation rate is generally smaller than the Arrhenius rate in mixing-controlled combustion and thus determines the overall reaction rates. It is noted in some gas combustion applications that the Eddy-Dissipation model over-predicts the flame temperatures and it should also note that this model is not appropriate to be used for multi-step reactions since a single reaction rate that is based on the turbulence dissipations is used for all steps.

When multi-step reactions are considered, the eddy-dissipation concept (EDC) model may be employed. EDC is an extension to the eddy-dissipation model that includes the detailed chemical mechanisms in turbulent flows. This model assumes that the reactions proceed according to the Arrhenius rates over the time scale which is defined as a function of the turbulence viscosity and dissipation rate. The Arrhenius rates may be calculated from the detailed chemical mechanisms employed for the turbulent reacting flows. However, it should note that reaction equations have to be numerically integrated time and again when the EDC is employed and it is computationally costly particularly when a large number of reactions are employed for a complex reaction flow.

The mixture fraction model approach has been widely used in coal combustion modelling applications and it is considered to be accurate over the finite rate eddy dissipation models since it can account for multi-species and multi-step reactions. Instead of solving explicitly the conservative equation for the species mass fractions, in the mixture fraction approach the mass fraction of each species, as well as all other thermo-chemical scalars, such as density and temperature, are uniquely related to the mixture fraction and enthalpy through a prescribed probability density function (PDF) of the mixture fraction, such as the delta and beta functions. This greatly reduces the number of conservative equations required to represent a combustion system and thus this saves the computational time.

In the mixture fraction model, when fast chemistry is assumed then an instantaneous reaction occurs as soon as the fuel and oxidant are simultaneously present at the same point. The mixture fraction f for the reaction can be defined by

$$f = \frac{\chi - \chi_O}{\chi_F - \chi_O} \tag{6.9}$$

$$\chi = m_F - \frac{m_O}{r} \tag{6.10}$$

where m is the mass fraction and the subscripts F and O refer to fuel and oxidant, respectively, r is the stoichiometric ratio of the oxidant and fuel by mass. The mixture fraction reflects the degree of mixing of fuel and oxidant in the flow. It can be proved that the mixture fraction f is equivalent to the mass fraction of the fuel in the system and thus it is a conserved quantity and the mean of the mixture fraction, $\bar{f}$, satisfies a conservative transport equation of the form:

$$\frac{\partial}{\partial x^i}\left(\rho u^i \bar{f}\right) = \frac{\partial}{\partial x^i}\left((\frac{\mu_t}{\sigma_t})\frac{\partial \bar{f}}{\partial x^i}\right) + S_f \tag{6.11}$$

where S_f is the source due to the transfer of mass into the gas phase from the combusting solid particles. In addition to the mean values of the mixing, in order to account for the instantaneous values due to the turbulence fluctuation, transport equation for the variance of the mixture fraction is usually solved. Further, chemical reactions are calculated using the appropriate fast reaction models and three key methods exist, namely, mixed-is-burned (or flame-sheet approximation), equilibrium chemistry, and the non-equilibrium chemistry (flamelet). The mixed-is-burned approximation is the simplest reaction scheme which assumes that the chemistry is infinitely fast and the fuel and oxidant do not coexist in space. Therefore, the species mass fractions can be determined directly from the instantaneous mixture fraction through reaction stoichiometry. However, this implies a complete single-step reaction to the final products and often results in a serious over-prediction of the peak flame temperature, especially in an oxygen rich environment. The equilibrium approach assumes that the reaction is fast enough for the local chemical equilibrium to always exist and thus the instantaneous species mass fractions can be calculated based on the minimization of the Gibbs free energy. Equilibrium calculations break down in situations where nonequilibrium prevails, such as in the rich side of flames. The flamelet model on the other hand calculates the instantaneous species mass fractions from the flamelets that have the same structure as laminar flames. The property of the laminar flames may be obtained by either experiments or calculations where a more realistic chemical kinetics effect can be incorporated. However, it should be noted that all the three approaches mentioned here are under the fast reaction assumption. For slow reactions with large reaction time scales, then finite rate chemistry models should be investigated.

When the mixture fraction approaches is employed in the modelling of co-combustion of two dissimilar fuels at least three streams of volatiles combustion need to be tracked separately. This makes the mixture fraction formulations much more complex and computational costly.

6.2   Modelling Pulverised Biomass Particle Combustion

Computer models for the heterogeneous combustion of coal have been relatively well developed because of the commercial needs of the power generation industry. However, models for biomass combustion have not been fully established and need further investigation. In general it is believed that in many cases the major processes of biomass combustion are similar to that of coal combustion. Therefore, modelling approaches that have been specially developed for coal combustion may be extended to the biomass combustion modelling incorporating biomass specific combustion rate constants in the CFD models.

Further, the use of biomass as a fuel presents a problem in that it is difficult to mill into small particles such as coal although treated biomass through torrefaction will improve its milling property. Consequently, pulverised biomass fuel usually contains particles with a larger size and aspect ratio than that of coal. This produces significant complexity in the CFD modelling of the particle aerodynamics and heat and mass transfer that need to be considered.

6.2.1 Particle Motion

In a typical pulverised fuel combustion system, the volumetric flow rate of the solid fuel is usually so small compared with that of the air. Therefore, the fuel particles in the flow stream are usually so far apart that the particles have little chance of interaction. In this situation the particle flow can be treated as dilute in the CFD model and a group of streams of particles over a range of sizes that represent the whole range of particle size distributions are tracked throughout the furnace and the different stages of the particle combustions are then modelled. In the commonly used framework of the Eulerian-Lagrangian modelling approach, the movement of the biomass particles are numerically tracked in the Lagrangian frame of reference and it is governed by the particle momentum equation obtained directly from Newton's second law as follows:

$$\frac{du_p^i}{dt} = F^i \tag{6.12}$$

The left hand side of Eq. 6.12 is the acceleration of the particle and on the right hand side are the forces that are exerted on the particle of unity mass. The forces which commonly act on a particle which is moving within the fluid flow are those which are produced by drag, gravity and, to a lesser extent, the gradients in the fluid velocity, pressure and temperature. For small particles that are suspended in a turbulent flow, additional forces, which are related to the energy which is consumed by putting the medium itself into motion, may be needed to be taken into consideration. There is still a considerable debate on the best way to calculate all these additional forces when modelling a particle that is suspended in a turbulent fluid flow.

In most modelling applications of pf coal combustions, the drag is often the only force that is considered and this is primarily because coal particles are usually small and have a high surface-volume ratio making the drag to be the dominant force over all other forces. The gravity force is also considered in many CFD calculations since it is the main reason for the production of the bottom ash. Since coal particles are not only small and but they tend to melt and become spherical during pyrolysis, therefore coal particles are typically treated as spherical particles in CFD models and thus the drag that acts on the particle by the gas phase can be expressed as follows:

$$F_D^i = C_D \frac{18\mu}{\rho_p d_p^2} \left(u^i - u_p^i \right) \tag{6.13}$$

where ρ_p, d_p and u_p are the density, diameter and the velocity of the particle, respectively. The drag coefficient, C_D, is often experimentally based and expressed as a function of the particle Reynolds number.

Modelling of the motion of biomass particles is much more problematic than that of coal particles. The size of the biomass particle used in the pf furnace is sometimes 10 times larger (in dimensions) than the coal particles and have a large aspect ratio, typically of the order of 10, and with a range of particle shapes such as cheap, disks, cylinders and even needles and fibres. These large biomass particles are prone to the effect of gravity and easily come down to the bottom ash in the furnace, particularly those particles that contain elements, such as the nodes and knots in wood, that are difficult-to-burn. Therefore, the effect of gravity in biomass combustion needs to be considered in the numerical simulations as it may have a significant effect on the model predicted particle trajectories in the furnace, although the effect may not be significant in a pf coal fired combustor with fine coal particles. Further, unlike coal particle that often softens and forms spherical particles during initial devolatilisation, the biomass particles tend to retain their initial form that is irregular in shape. Therefore, it will be inaccurate to take biomass particles as spherical particles as we do for the coal particles in the CFD model, since the shape of the particle has a strong influence on the path of the particle motion in the combustion chamber. There are different ways of dealing with the effect of the particle shape on the particle tracking and in principle this is achieved by considering the aerodynamic effects of the irregularity of the particle with more or less reference to a spherical particle. One way of dealing with the irregularity in the particle shape is to employ the so-called aerodynamic diameter of the particle, which is defined as the diameter of an equivalent sphere which has a unit physical density and the same settling velocity as that of the particle in question. More commonly used ways of modelling the motion of irregular particles is the introduction of a particle shape factor, or sphericity, which is defined as the ratio of the surface area of an equivalent spherical particle, s, and the real particle surface area, S, for example, as follows:

$$\phi = \frac{s}{S} \tag{6.14}$$

Then the drag that acts on a non-spherical particle may be formulated as a function of particle Reynolds number based on experimental data as follows:

$$C_D = \left(1 + a_1 \text{Re}^{a_2}\right) + \frac{\text{Re}}{24} \frac{a_3 \text{Re}}{a_4 + \text{Re}} \tag{6.15}$$

where a_1, a_2, a_3 and a_4 are empirical constants and often expressed as a function of the particle shape factor. The diameter of the sphere of equal volume has been used in the drag and particle Reynolds number calculations. There are numerous other ways of calculating the drag coefficient. Inevitably, there is debate on the accuracy of the formulations in modelling the drag of a vast variety of non-spherical particle shapes and aspect ratios because experiments can only be performed with a limited number of shapes and size of particles. The introduction of the shape factor to some extent represents the effect of the aspect ratio of the particle on the motion of the particle but the details of the particle are not calculated exactly. For disk and chip shaped particles, the particles may be tumbling during their penetration of the highly turbulent gas flow in the furnace, and the drag that is exerted on the particle is significantly influenced by the instantaneous orientation of the particle. However, the instantaneous orientation of a particle is influenced by the initial state of the particle as it enters the computational domain which itself is arbitrary and difficult to determine. More detailed aerodynamics of the flow over irregular shaped object is very complex and is impractical to be modelled in detail. For the purpose of the computational simulation of a group of dispersed fuel particles of vast diversity of shape and size, the practical way of effectively calculating the drag acting on the particles is to use experimental based empirical formulae.

Fig. 6.2 Model predicted biomass particle trajectories in the EON 1 MW combustion test furnace. The initial equivalence particle size is 0.75 mm with a shape factor of 0.1. The scale is *coloured* by particle mass (kg). The EON Test Facility is at Ratcliffe, UK

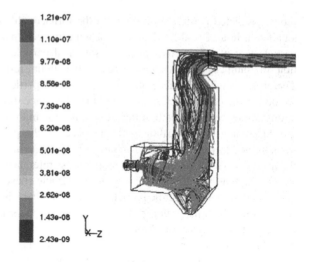

It is seen in Fig. 6.2 that most of the particles with a shape factor of 0.1 penetrate to the fly ash but some particles do comes down to the bottom of the furnace. However, if the particles were spherical coal particles then all the particles of this size would be expected to end up in the bottom ash. Model predictions show that on decreasing the biomass particle shape factor, i.e. the particle is more non-spherical, then the biomass bottom ash decreases and the fly ash increases.

For particles that are suspended in a fluid flow, in addition to the drag and the gravity, the forces that act on the particles may also include those that are produced by the gradients in fluid velocity, pressure and temperature in the undisturbed fluid flow. Particles traverse in a fluid with a velocity gradient may experience a lift force if the particle is rotating. The pressure difference in the ambient fluid stream also results in a force that acts on the particle and this may become significant when the size of the particle is large. A temperature gradient in the gas flow may also impose a force on the suspended particle, known as the thermophoretic force, in the direction opposite to that of the temperature gradient. This may be significant near to a cooled or heated surface, in particular for submicron particles. However, in most situations, these forces are small for larger particles and can be ignored in most mathematical models. The drag and the gravity forces are the dominant forces that control the motion of the biomass particles in the combustion furnace.

Large particle sizes and aspect ratios also present a number of heat and mass transfer problems for the CFD modelling. For a pulverized coal particle, of say, typically about 50 μm, the particles can be heated up instantaneously. The pyrolysis of such small particles is essentially unaffected by the heat transfer within the particle. Therefore, in the CFD model, the thermal state of a particle can be represented by a single uniform particle temperature. However, for large particles, say 1 mm or larger, a temperature gradient from the surface to the centre of the particle will build up. The thermal and mass transport limitations and the temperature gradients within the particle will have an effect over the life time of the particle as it passes through the furnace. At present, most CFD models ignores the effects of heat and mass transfer within the particle and this is acceptable in coal combustion simulations, as the temperature is fairly uniform inside the small pulverized coal particles and the mass transfer may be so fast so that the carbonization uniformly proceeds throughout the particle. However, the effect of heat transfer within relatively large biomass particles must be accounted for in the CFD simulations.

In order to investigate the instantaneous temperature distribution inside large size and high aspect ratio biomass particles, CFD models may be developed where the heating up process of the particles may be numerically calculated. Figure 6.3 shows the model predicted transient temperatures on the surface and at the centre of the two cylindrical wood particles when they are heated up in a nitrogen gas stream. The two particles have the same diameter, i.e. 9.6 mm, different aspect ratios of 1.0 and 4.0 and moisture content of 6 %. Figure 6.3a, b presents the comparisons of the numerical predictions of the wood particle surface and centreline temperatures with the experimental data. Although the particle surface temperature and centre temperature reach the same value at the end of pyrolysis, larger particles will take a longer time to heat up and have a greater internal temperature gradient. Similar situations exist for biomass fuel particles in pf furnaces. This slow heating up process will significantly affect the pyrolysis, as well as the devolatilisation, of large biomass particles in the combustion, and this may have a significant impact on the flame structure and char burn-out. In addition, for large particles, devolatilisation and char combustion may coexist in different parts of the particle. This is particularly true for irregular shaped particles. However, at present, it is very difficult to model the combustion of such irregular shaped particles in detail in a full scale combustion simulation. Although substantial progress has been made in recent years, accurate sub-models to account for the effect of particle shape on the heat and mass transfer within the particles have yet to be developed and they require a significant amount of effort by both experimentalists and CFD modellers.

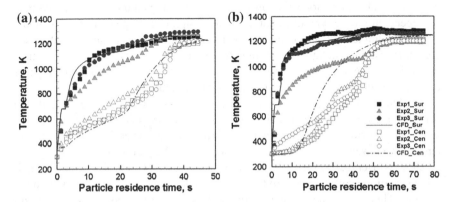

Fig. 6.3 Temporal variations of the surface and centreline temperatures of wood particles of dp = 9.6 mm and moisture content of 6 % (wb) during pyrolysis in nitrogen. **a** Particle aspect ratio 1.0, and **b** particle aspect ratio 4.0

6.2.2 Heat Transfer

Once the particles emerge from the burner, they are heated up by a combination of convective and radiative heat transfer from the furnace walls and hot gases and continuously absorb heat during the processes of volatiles and char combustion. In the CFD model, the heat lost or gained by the particle as it traverses the furnace will be presented as a source or sink of heat in the energy equation for the gas phase.

The instantaneous temperature of a particle travelling in the combustion furnace may be calculated as follows:

$$m_p c_p \frac{dT_p}{dt} = hA_p\left(T - T_p\right) + A_p \varepsilon_p \sigma \left(T_{rad}^4 - T_p^4\right) - Q_{dry} + Q_{reac} \quad (6.16)$$

where m_p is the mass of the particle, c_p is the heat capacity of the particle, A_p is the surface area of the particle, T is the local gas temperature, T_p is the particle temperature, h is the convective heat transfer coefficient, ε_p is the particle emissivity, σ is the Stefan-Boltzmann constant, and T_{rad} is the radiation temperature that is a function of the radiation intensity.

Radiative heat transfer is one of the dominant processes for heating up fuel particles in a large furnace. There are a number of radiation models available for combustion modelling. However, in pf combustion modelling applications, the P-1 model and the Discrete Ordinate model are often employed. The P-1 model is a first-order approximation to the radiation transfer equations by integration where the radiation intensity is represented by a series of spherical harmonics. The Discrete Ordinates radiation model is an approximator for the radiative transfer equation by discretizing the entire solid angle (4π) with a prescribed finite number of solid angles. On increasing the fineness of the angular discretization the accuracy of the model calculations may be improved with an additional cost of computation. There is debate on the use of radiation models in pulverised biomass and coal combustion simultaneously and the temperature field obtained may be dependent on the type of radiation model employed. In general, the P1 model is easy to implement and computationally robust. It is accurate for pulverised coal combustion simulations with radiation scattering, in particular if the optical dimension of the medium is large. The Discrete Ordinates radiation model is more computationally expensive but potentially more accurate.

In addition, as discussed in the previous section, for the usually very small coal particles, there would be a very limited temperature gradient within the particle in a very short period of time and the entire particle can be virtually heated up instantaneously. However, when the size of biomass particles can be up to 20–200 times as large as average pulverized coal particles of say, ~50 μm. Then the thermal-resistance of the particle may produce a noticeable temperature gradient within the particle. This delays the heating up of the biomass particle as a whole and will have a further effect on the subsequent processes of drying, devolatilisation and char combustion. Therefore, a delay in the ignition of the biomass particle can be observed. The stability of the flame will also be influenced by the rate of the heating up of the particles, in addition to the heating values of the fuel. Therefore, for a thermally thick large biomass particle, where there exists a significant temperature gradient within the particle, then additional heat balance models within the particle have to be developed in which both thermal-physical properties and the shape and dimensions of the particle may be considered.

6.2.3 Devolatilisation

The process of devolatilisation was discussed earlier in Sect. 3.3. In biomass CFD models it plays a very important role because of the large quantity of volatiles produced. In most coal models, the process of drying can be incorporated in the devolatilisation model. However, in biomass combustion, the water content is usually larger and in some instances it may dominate the combustion process. Consequently, the drying step is often treated as a separate process as set out earlier in Eq. 3.1.

The rate of volatile release determines the volatile mass transfer from the solid particle to the gas phase and the subsequent volatile gas combustion. As previously described there may be catalytic effects which complicate the process. The input data are the reaction kinetics for the particular biomass using experimental data or information from one of the network codes described earlier. At present there is a considerable amount of data suitable for fixed bed combustion but not much available for higher temperature pf furnaces. Network codes may be used and whilst they may be applicable to commonly used biomass the range covered at present is not large. This is in contrast to the wealth of information available on coal because of long established commercial interests of the coal industry.

The other inputs required are usually the proximate and ultimate analyses and the high temperature volatile yield which is higher than that given by the Proximate Analysis. This value can be obtained from the Network Codes or determined experimentally. Then the rates and yields of all major species, molecular weight distributions and elemental compositions of tar and char from the fuel undergoing devolatilisation can be predicted and used for CFD model inputs. Thermal data such as specific heats as well as densities are also required for the heating-up models.

The devolatilisation process for a particular biomass may be modelled in the CFD model with different levels of complexity. The simplest model is to assume a constant rate of devolatilisation, the single kinetic reaction model, where typically representative values for both the devolatilisation rate and devolatilisation temperature may be used so that the overall process of the volatile release can be modelled. In this situation, the rate of volatile release may be assumed to be first-order dependent on the amount of volatiles remaining in the particle as previously described in Chap. 3. However for accurate modelling of the lower temperature processes such as ignition or the combustion of thermally thick biomass then the use of a three stage devolatilisation model is the best approach.

6.2.4 Char Combustion

The main features of char combustion were considered in Chap. 3. Char combustion occurs principally after the end of the devolatilisation, although in practice a small fraction of char oxidation begins earlier and co-exists with the devolatilistion step. Char combustion is a much slower process and therefore it determines the extent of burn-out of the solid fuel in the furnace; this is experimentally measured as the carbon-in-ash, i.e. the carbon remaining in the ash at the exit of the combustor. In typical large-scale pulverised fuel plants the furnace residence time is of the order of several seconds to allow for complete char burnout. Similar furnaces are used for the combustion of biomass and although the char accounts for a much smaller fraction of biomass fuels, firing large biomass particles in the existing pulverized coal furnace may raise concerns over unburned carbon.

Similar factors apply to combustion in fixed, travelling or fluidised beds. That is, the overlapping devolatilisation and char burnout stages and the concern about unburned carbon remaining in the ash. An extra concern in the combustion of fixed bed combustion is the inhibiting effects of ash which can form a layer on the outside of the char particle. In pf combustion the motion of the particle may remove much of this ash and this is the case in travelling and fluidised bed combustion too. In fixed beds with for example this has to be taken into account (Mehrabian et al. 2014).

The critical step is the calculation of the rate at which solid char particles is oxidised. Coal char combustion can be modelled relatively accurately but this is not yet the case of biomass. The burnout rate is more complicated as it is affected not only by the composition of the biomass fuel but also by the shape and size of the particles and therefore the heating-up process as previously outlined. A further complication is that biomass used in real world reactors generally consists of mixtures of woods and may contain bark, roots and impurities such as soil. The milled biomass is not uniform chemically and in the case of wood it contains 'knots' which have a higher lignin content and are thus less reactive. This is also true for 'nodes' in Miscanthus and 'knees' in straw.

There is a paucity of reactivity data available for biomass char at high temperatures and is often conflicting. The combustible portion of both coal and biomass char particles is basically the solid carbon. The abundant inorganic elements in the biomass char particles tend to remain within the ash. As soon as the fixed carbon on the surface of the char particles is molecularly in contact with the oxidant from the gas phase then the reaction proceeds and produces the gas phase products. At the high temperatures that typically occur in the pf combustion furnace, diffusion is more likely to be the control mechanism of the entire char combustion than the chemistry. For biomass particles, large aspect ratios are usually a typical feature and any shape from plates, chips or round cylinders to needles and fibres may be present. The immediate effect of this is the increased surface area of the particle compared with the equivalent spherical particle and this would enhance the diffusion of the oxidant to the pores within the biomass char particles.

Metal and inorganic elements account for a fairly large percentage of char mass and thus may effect on the combustion rate of biomass chars. There is a considerable uncertainty about the role of catalytic metals in char combustion. Generally it is believed that at high temperature char combustion in pf furnaces, where a very fast chemistry exists, the catalytic effect is not as significant as it is in low temperature combustion. Further, biomass chars contain high levels of oxygen and low levels of hydrogen which tend to prevent graphitic structures from developing in biomass chars.

6.3 Modelling Pulverised Fuel Co-firing in Power Stations

Co-firing coal and biomass in a pf furnace involves the simultaneous combustion of a blend of two very different fuels in terms not only of their different chemical compositions and reactivity during combustion, but also different particle aerodynamics and heat and mass transfer. The combustion behaviours of two dissimilar types of solid fuels have significant nonlinear characteristics. When co-firing coal and biomass in a pf coal furnace, the vast differences in the combustion properties of biomass have a noticeable impact on the combustion of coal, see for example, Kaer et al. (1998) and van Loo and Koppejan (2008).

In CFD simulations, pf solid fuel combustion is usually modelled in the Eulerian-Lagrangian framework where the motion of particles is tracked individually in a Lagrange frame of reference. When two very distinct fuels are fired simultaneously, then two types of volatiles with different compositions are produced and two types of char particles are formed. Therefore, in the co-firing modelling, up to four different fuel streams need to be tracked in order to model the different processes of drying, devolatilisation, volatile combustion and char combustion in the furnace with precision and, in particular, capture the nonlinear interactions of coal and biomass combustion processes. This put significant demand on both the mathematical formulations and computational techniques.

It would be more or less straightforward to model an additional fuel's combustion by using the eddy break up type of models, such as the Eddy-Dissipation turbulence-chemistry interaction model for volatile combustions. In this situation, the additional biomass volatile and char combustions can be modelled by putting additional source terms into the appropriate conservation equations, such as the energy equation, as well as in the mass and chemical species conservative equations. Clearly, the composition of the biomass volatiles, biomass volatile oxidation reaction mechanisms and the heating values have to be prescribed and this, however, also involves significant uncertainty. With the Eddy-Dissipation model, both volatile and char combustion are modelled as global reactions and therefore intermediate reactions and species could not be calculated.

The mixture fraction approach is also used to model the co-firing of different solid fuels, which is believed to give more accurate predictions of the turbulent gas combustions than does the Eddy-Dissipation type of models. However, the mathematical formulations with the multi-mixture fraction approach would become very complex for co-firing two distinctive solid fuels and this would cause problems in the numerical solution techniques.

There have been a number of attempts to model the blends of coal and biomass combustion using the multiple mixture fraction approach, such as Yin et al. (2010), Abbas et al. (1994), Dhanaplan et al. (1997) and Sami et al. (2001). Various simplifying assumptions have to be made to reduce the level of complexity of the computations. In a simple case, for example in the case of two fuels and the oxidant is shared amongst them, then the mixture fractions can be defined by:

$$f_i = m_{F,i} + \frac{m_{P,i}}{1 + r_i}, \quad i = 1, 2 \tag{6.17}$$

where m is the mass fraction and the subscripts F, P and O refer to the fuel, products and oxidant respectively. The mean of both mixture fractions Fi, plus their variance, satisfy the standard transport equation for a conserved quantity. Then the equations for the mass fractions of fuel may be solved and the oxidant is shared between the two fuels. Then the mass fractions of the oxidant and products may be obtained from:

$$m_{P,i} = (1 + r_i)(f_i - m_{F,i}) \tag{6.18}$$

$$m_O = 1 - m_{F,1} - m_{F,2} - m_{P,1} - m_{P,2} \tag{6.19}$$

Some detailed formulations for two stream mixture fraction calculations may be found in, for example, the commercially available CFD software ANSYS FLUENT manual. It can be expected that the formulations for three-stream mixture fractions, which are required in order to treat two different off-gases separately for a co-firing application, will become extremely complex and the solution of the mixture fraction equations, the setup of the lookup table and the data retrieval will be much more complex and computationally time consuming.

There are a number of commercially available CFD codes that model coal combustion and NO_x formation with reasonable accuracy. However, handling of two different solid fuels is not yet a typical feature of the commercially available CFD software codes and there are very few full-scale CFD simulations of co-firing in the open literature. This, to some extent, reflects the difficulties in modelling the co-firing of different fuels.

Within the framework of the Eulerian-Lagrangian approach, Backreedy et al. (2005) performed a full three-dimensional CFD modeling of co-firing pine wood and bituminous coal in a 1 MW industrial combustion test facility. The general approach taken was to assume that many aspects of the coal and biomass combustion processes are common, and that the key sub-models used, such as those applicable to drying, devolatilisation, volatile and char combustion, are the same as for coal with different model constants being applied to the two fuels and a kinetic-diffusion model for the biomass char particles. The particle shape factor approach has been used to calculate the effect of the non-spherical shape of the biomass particles on the drag that acts on the particles. The eddy dissipation model was employed for the turbulence chemistry interaction and the RNG model for turbulent swirling flow. The coal char combustion model was based on Smith's model (described in Chap. 3) and this was modified to allow for the effects of macerals, variable surface area and annealing on the burnout of the char, described by Backreedy et al. (2005) and references therein. The P1 model for radiation and the thermal/prompt/fuel mechanisms NO_x have been employed. For biomass combustion, the formation of biomass volatiles and char are calculated by using the FG-Biomass network programme in order to estimate the amount and rates of the volatile release. In this particular study, a smaller amount of biomass has been used in the co-firing, 3 % by mass, and the work focused on the effect of the particle shape and size, volatile release rate and the char burnout rate on the unburned carbon in biomass ash. The study showed that, for the pine wood investigated, small particles, less than about 200 μm, burn out completely. However, the burnout of large biomass particles depends on both the size and shape of the particle and burning rates.

Gera et al. (2002) performed a simulation for the co-firing of coal and biomass switchgrass in a utility T-fired 150 MWe boiler and 10 % (by heat) switchgrass and 90 % coal was used. Again, an Eulerian-Lagrangian approach is used. The simulation uses the standard k-ε turbulence model for the turbulent fluid flow and the eddy dissipation model for the turbulence-chemistry interaction. The radiation model employed is the Discrete Ordinate model. For the biomass char combustion, a diffusion controlled rate model was assumed taking into considerations of the effect of large aspect ratios of biomass particles on carbon burnout where temperature gradient inside the biomass particles is also considered. There are some other similar modelling papers on co-firing which employed different CFD codes, such as Abbas et al. (1994), Dhanaplan et al. (1997) and Yin et al. (2010). The simulations were primarily based on relatively well developed coal combustion sub-models with some modifications and/or further developments in order to make them applicable to the unique features of biomass fuels in terms of both chemical reactions and particle aerodynamics.

Since biomass usually has a lower heating value therefore more biomass fuel will be required in order to keep the same thermal input. This increase in the fuel flow rate

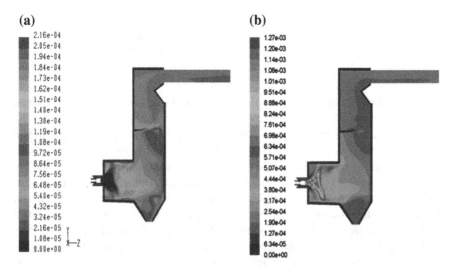

Fig. 6.4 Computed. **a** NO and **b** HCN profiles-same reactor as Fig. 6.2

Fig. 6.5 An example of modelling combustion in a power station boiler using coal or biomass as the fuel, with different levels of oxygen. Temperature distribution **a** air–coal, **b** air–biomass, **c** oxy25 %-coal, **d** oxy25 %-biomass, **e** oxy30 %-coal, and **f** oxy30 %-biomass (Black et al. 2013, with permission)

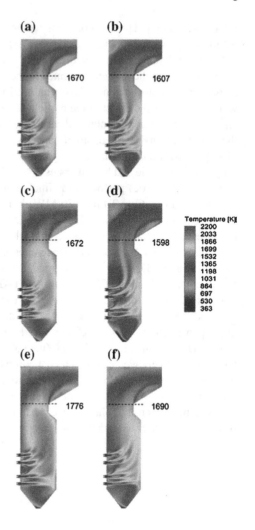

may cause the flame to move away from the burner mouth, thereby creating a flame stability problem and cause higher emission levels. If pure biomass is used then the whole combustion processes or the combustor itself may need to be changed to suit an efficient combustion of the fuel with, for example, different air splits in the burner.

The modelling of the combustion of small pulverised biomass particles is readily undertaken using CFD methods, although the modelling of the release of the pollutants is much harder. Many studies have been made of pf coal combustion and in a similar way; studies have been made of co-firing using a two-component fuel. Since the amount of biomass is often less than 20 % thermal the main flame properties such as temperature or velocity are largely determined by the coal combustion processes. The key to the correct prediction of NO_x is the correct prediction of the temperature distribution in the furnace especially if staged burners are used.

In large pf-fired, multi-burner furnaces, which use 20 % or more biomass over-all, the pulverised biomass is injected via individual biomass burners using 100 % biomass. There have been only a few such studies. Ma et al. (2007) have studied the combustion of pulverised wood and determined NO_x and unburned carbon in ash in a 1 MW Combustion Test Facility. Typical computed profiles for NO_x and the intermediate HCN are shown in Fig. 6.4.

The amount of NO computed on the basis that the fuel-N converts to HCN and then reacts to NO using the ANSYSY FLUENT kinetics is 89 ppm, whereas if the NH_3 mechanism is used it is 266 ppm, the former being closer to the experimental results. In the case of most other fuels it is postulated that the latter is the main route, although it has been observed in industry that the injection of biomass in co-firing does not produce the expected reduction in NO associated with ammonia injection.

An example of modelling combustion in a power station boiler using coal or biomass as the fuel and different levels of oxygen is shown in Fig. 6.5. Temperature distribution (a) air–coal, (b) air–biomass, (c) oxy25-coal, (d) oxy25-biomass, (e) oxy30-coal, and (f) oxy30-biomass.

6.4 Modelling Fixed Bed Combustion

Much of the early work on modelling pollutant formation from grate combustor has been summarized by Saastamoinen et al. (2000). There have been significant advances since then in the chemistry and physics of modelling such combustors and these have been recently described by Mehrabian et al. (2012, 2014). In general modelling presents a number of problems because, whilst the gaseous combustion occurring in the furnace chamber above the surface of the bed is relatively easy to simulate, this is not true of the bed itself. The bed consists of irregular shaped particles where the initial fuel size may be 10–100 mm more, and whilst the term 'fixed' is used, in a real system the bed particles are constantly shrinking in size and the bed is moving. This is clear from Fig. 2.2. The fuel is often fed on a batch basis and in addition the bed is not fixed in space and holes can appear through the bed in a random manner. Moving or travelling grate furnace arrangements are simpler in many respects in that input and movement of the fuel particles in the bed can be treated more easily in a statistical manner, although channelling still occurs. Many of the modelling studies undertaken relate to simplified systems and directed to increasing furnace efficiency which can directly reduce pollutant formation, but few modelling studies have been concerned directly with pollution formation. The introduction of wood pellet burners in the last decade has simplified these calculations since the fuel is more uniform and combustion much more consistent. Many researchers have carried out simulation studies of biomass combustion for large particles and pellets as discussed in this chapter.

Some modelling studies have been made of pollutants formed from small village cook stoves. Bhaskar Dixit et al. (2006) and Kausley and Pandit (2010) studied flow and heat transfer in cook stoves but not pollutant formation. Ndiema et al. (1998)

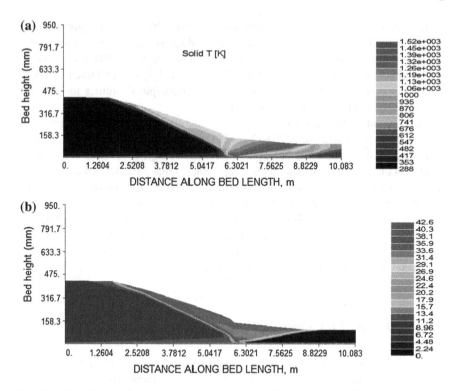

Fig. 6.6 Modelling moving grate combustion showing. **a** Temperature profile inside a grate bed the temperature of the solid bed is in K. **b** Carbon profile inside a grate bed (Yang 2014, Personal Communication, Biomass Power Ltd., UK)

investigated the formation of NO_x, methane and smoke experimentally in a 1-D plug flow model of a cook stove, and the gaseous pollutants followed by Chemkin (2014) modelling. Here the factors determining the methane emission were examined.

A number of studies have been made of larger fixed bed combustors for residential and commercial applications combustion using a variety of fuels (Galgano et al. 2006; Huttunen et al. 2006a, b). The earlier published modelling studies were related to the use of fuels such as wood with an irregular size or briquettes made of sawdust or straw. Later studies were much more general and looked at the influence of the many factors involved such as bed size, level of moisture, particle size etc. Cylindrical pot furnace studies are useful because of the consistency of the combustion process, and this has been used to investigate NO_x formation from several woods and wood pellets. Because of intermittent combustion in practical systems, the complex gas flow, heat transfer and reaction within the bed, many models have been concerned efficiency studies or partial modelling of pollutant formation in the combusting gases above the bed. The introduction of pellet burners in the last decade has not only improved combustion but simplified these calculations since the fuel is more uniform and the feeding process and combustion much more consistent. Recently researchers have carried out simulation studies of biomass combustion for large particles and pellets and calculated NO formation (Klason and Bai 2007).

Modelling the emission of pollutants from larger thermal capacity (about tens of MW) grate units follows the same modelling methods, although the gas flows, especially in the injection of secondary air and in the heat exchanger section, are more complex. CFD modelling in these furnaces has been extensively reviewed by Yin et al. (2008). This review shows how little work has been devoted to modelling pollutant formation in biomass combustion. There are a number of examples related to both fixed bed and travelling grate combustion. Yang et al. (2005) simulated packed-bed combustion of biomass particles with size ranging from 5 to 35 mm. A typical computation for a travelling grate is given in Fig. 6.6.

Whilst most of the studies have been concerned with the major fuels, of wood and straw, some studies have also been made of other biomass materials such as bagasse.

6.5 Modelling Fluidised Bed Combustion

Fluid bed combustion is one of the major methods of utilizing biomass for power generation. A considerable number of plants are installed using a variety of different types of biomass, being particularly suitable for wet agricultural residues. The designs are largely based on those for coal combustion and two types of designs are used, namely bubbling fluidized bed (BFB) and circulating fluidized bed combustors (CFBC), both of which operate with bed temperatures of about 800 °C. The main difference between coal combustion and biomass combustion are the large amount of volatiles produced in the latter case, the greater reactivity, and lower SO_x and NO_x (and N_2O) emissions. Thus the BFB system is suitable for many applications, where neither staged combustion nor in-bed sulphur removal are necessary. It also has the advantage that large particle sizes are possible and less prone to bed agglomeration due to the alkali metals in the ash. Typical sizes are 50–100 MW and, as such, are capable of using major pollution control equipment. Large plant of this type usually has to conform to National Emission Directives and so the level of emission has to be achieved by a combination of the combustion chamber design and the flue gas clean-up plant, which will involve particulate and possibly NO_x units. If the K, Cl, Si contents are high, this can lead to submicron emissions. As with most large-scale combustors the amount of NO_x and the metals can be determined by equilibrium modelling.

6.6 Modelling Pollutant Emissions

6.6.1 Nitrogen Oxides

The modelling of biomass has followed the methods used for coal combustion. In the latter case, considerable attention has been directed to the combustion of pulverised coal, which is the major industrial use of coal, and much less attention directed to fluidised bed combustion and fixed bed combustion. This is the same order of difficulty in modelling. In modelling pollutant formation much more attention has been directed to the gases, NO and CO than to the particulates. Here the outcomes of some major examples of modelling of pollutant formation will be considered.

Similar to coals biomass contains nitrogen although may in a different level and form. When biomass is burnt it produce nitrogen oxides but because furnace temperatures are kept fairly low to prevent slagging the production of thermal-NO_x is low. The fuel-nitrogen is mainly released into the volatiles. Volatile combustion is usually considered as a mixing controlled fast reaction and thus the detailed reaction mechanisms can usually be ignored and global one step or two step kinetics are usually employed. However, in the situation where slow reactions and intermediate species dominant, such as NO_x and SO_x reactions, then finite reaction schemes have to be considered.

There are many studies on NO_x formations for coal combustions, but most are focused on thermal and prompt formation routes which are the key pathways of NO_x formation in high temperature combustions, such as coal and nature gas flame. For biomass combustion occurring at a relative low temperature, the fuel bound nitrogen will be the focus of the investigation of the NO_x formation. The fuel bound nitrogen is typically released in the form of either HCN or NH_3 during combustion before being transformed into N_2 or NO. The exact fraction of each intermediate species formed remains the subject of ongoing research, and this will affect the model prediction of the overall NO_x formation. However, it is generally accepted that ammonia is the principal product of fuel-nitrogen conversion in biomass fuel stocks which may be orders of magnitude higher than those of other fuel-nitrogen species, for example see Zhou et al. (2005). At the moment, quantitative modelling of NO_x formation in biomass combustion is still at an early stage of research.

6.6.2 SO_X Emissions

The sulphur content in the biomass is small; therefore combustion of biomass fuel itself has not become a big issue in SO_x formation except in relation to fine particulate formation where even a small amount of S may be significant. The formation of acid rain is not really an issue with biomass firing although the global effects on the climate are significant. The major concern is during co-firing with coal in large electricity generation units where SO_x up to the permitted limits may be emitted. However co-firing biomass with coal in normally 100 % coal fired plants permits a reduction in the overall SO_x emissions which may result in desulphurisation plant running at a reduced load or not being needed.

6.6.3 Modelling Aerosol Pollutants

Aerosols generated from biomass combustion can be a result of an incomplete combustion such as soot, polyaromatic hydrocarbons (PAH) as well as fragments of ash (Eriksson et al. 2014). However, majority of the aerosols are produced as a result of alkaline metal and chlorine presents in the biomass which lead to the formation of sticky and corrosive components and submicron particulates, such as KCl and K_2SO_4. As discussed before, the form in which the metal species are released from combustion of biomass depends on the relative contents of the potassium and chloride in the biomass. Once released as a part of the volatile, and the oxidation gaseous of char to a less extent, the potassium and chlorine undergo a series of chemical and physical transformation steps to form more stable KCl and K_2SO_4 compounds in gaseous phase and then through nucleation and condensation to form aerosols. The detailed kinetics of the sulphation of alkali spices involves hundreds of intermediate reactions, which are too complicated for use in computational fluid dynamics calculations (Garba et al. 2012). Reduced reaction mechanisms are usually employed and the major reactions include (Glarborg and Marshall 2005):

$$KCl + H_2O = KOH + HCl \tag{R 6.1}$$

$$KCl + SO_3 = KSO_3Cl \tag{R 6.2}$$

$$KSO_3Cl + H_2O \rightarrow KHSO_4 + HCl \tag{R 6.3}$$

$$KOH + SO_3 \rightarrow KHSO_4 \tag{R 6.4}$$

$$KHSO_4 + KCl \rightarrow K_2SO_4 + HCl \tag{R 6.5}$$

The concentration and transportation of the aerosols from KCl and K_2SO_4, as well as other particulates e.g. soot particles, can be modelled using the standard transport equations in the form such as (Jöller et al. 2007):

$$\frac{\partial \rho \phi}{\partial t} + \frac{\partial \left(\rho u^i \phi \right)}{\partial x^i} = \frac{\partial}{\partial x^i} \left(\Gamma \frac{\partial \phi}{\partial x^i} \right) + S_\phi \tag{6.20}$$

where φ is the mass fraction of the aerosol, Γ is the diffusivity and S_ϕ is the production of the aerosol.

The aerosol formation and growth in a combustion system may through both homogenous and heterogeneous nucleation and condensation. Based on classic kinetic theory, homogenous nucleation rate may be given in the form (Hagen and Kassner 1984)

$$J = C^* F A^* a Z \tag{6.21}$$

where C^* is the concentration and A^* is the surface area of critical sized nucleation embryos, F is the flux of molecules to the cluster surface, a is the sticking coefficient, and Z is the Zeldovich factor which accounts for non-equilibrium effects in the embryo concentration. The condensation rate of the species to a particle is a function of vapour pressure, p and saturate pressure $p_{sat,p}$ at the particle surface temperature, typically expressed in the form:

$$m = c \frac{(\dot{p} - p_{sat,p})}{RT} \tag{6.22}$$

where c is mass transfer coefficient, R is the gas constant and T is the gas temperature.

Coagulation of particles due to particle collisions among each other may also be considered. The coagulation process would not influence the total concentration of the particles but does have an impact on the particle size distributions and particle composition. Thus nucleation, coagulation and condensation of alkali and chlorine species are responsible for the formation of majority of the submicron aerosol emissions from biomass combustion. The concentrations of KCl and K_2SO_4 in the flue gas of a power generation boiler also have significant impact on the fouling potentials on the heat exchangers.

References

Abbas T, Costen P, Kandamby NH, Lockwood FC, Ou JJ (1994) The influence of burner injection mode on pulverized coal and biomass cofired flames. Combust Flame 99:617–625

Andersson B, Andersson R, Hakansson LH, Mortensen M, Sudiyo R, van Wachem B (2012) Computational fluid dynamics for engineers. Cambridge University Press, Cambridge

ANSYS FLUENT (2014) ANSYS Inc., USA

Backreedy RI, Fletcher L, Jones JM, Ma L, Pourkashanian M, Williams A (2005) Co-firing pulverised coal and biomass: a modelling approach. Proc Combust Inst 30:2955–2964

Bhaskar Dixit CS, Paul PJ, Mukunda HS (2006) Part I: experimental studies on a pulverised fuel stove. Biomass Bioenergy 30:673–683

Black S, Szuhánszki J, Pranzitelli A, Ma L, Stanger PJ, Ingham DB, Pourkashanian M (2013) Effects of firing coal and biomass under oxy-fuel conditions in a power plant boiler using CFD modelling. Fuel 113:780–786

Chemkin (2014) Reaction design. www.Reactiondesign.com

Chung TJ (2002) Computational fluid dynamics. Cambridge University Press, Cambridge

CRECK (2014) http://creckmodeling.chem.polimi.it

Dhanaplan S, Annamalai K, Daripa P (1997) Turbulent combustion modelling coal: biosolid blends in a swirl burner. Energy Week vol IV, ETCE, ASME, Jan 1997, pp 415–423

Eriksson AC, Nordin EZ, Nyström R, Pettersson E, Swietlicki E, Bergvall C, Westerholm R, Boman C, Pagels JH (2014) Particulate PAH emissions from residential biomass combustion: time-resolved analysis with aerosol mass spectrometry. Environ Sci Technol 48:7143–7150

Galgano A, Di Blasi C, Horvat A, Sinai Y (2006) Experimental validation of a coupled solid- and gas-phase model for combustion and gasification of wood logs. Energy Fuels 20:2223–2232

Garba MU, Ingham DB, Ma L, Porter RTJ, Pourkashanian M, Williams A, Tan HZ (2012) Prediction of potassium chloride sulfation and its effect on deposition in biomass-fired boilers. Energy Fuels 26:6501–6508

Gera D, Mathur MP, Freeman MC, Robinson A (2002) Effect of large aspect ratio of biomass particles on carbon burnout in a utility boiler. Energy Fuels 16:1523–1532

Glarborg P, Marshall P (2005) Mechanism and modeling of the formation of gaseous alkali sulfates. Combust Flame 141:22–39

Hagen DE, Kassner JL Jr (1984) Homogeneous nucleation rate for water. J Chem Phys 81:1416–1418

Huttunen M, Saastamoinen J, Kilpinen P, Kjaldman L, Oravainen H, Bostrom S (2006a) Emission formation during wood log combustion in fireplaces—part I: volatile combustion stage. Prog Comput Fluid Dyn 6:200–208

Huttunen M, Saastamoinen J, Kilpinen P, Kjaldman L, Oravainen H, Bostrom S (2006b) Emission formation during wood log combustion in fireplaces—part II: char combustion stage. Prog Comput Fluid Dyn 6:209–216

Jöller M, Brunner T, Obernberger I (2007) Modeling of aerosol formation during biomass combustion for various furnace and boiler types. Fuel Process Technol 88:1136–1147

Kaer SK, Rosendahl L, Overgaard P (1998) Numerical analysis of co-firing coal and straw. In: Proceedings of the 4th European CFD conference, Athens, Greece, 7–11 Sept 1998, 1194–1199

Kausley SB, Pandit AB (2010) Modelling of solid fuel stoves. Fuel 89:782–791

Klason T, Bai XS (2007) Computational study of the combustion process and NO formation in a small-scale wood pellet furnace. Fuel 86:1465–1474

Ma L, Jones JM, Pourkashanian M, Williams A (2007) Modelling the combustion of pulverized biomass in an industrial combustion test furnace. Fuel 86:1959–1965

Mehrabian R, Zahirovic S, Scharler R, Obernberger I, Kleditzsch S, Wirtz S, Scherer V, Lu H, Baxter LL (2012) A CFD model for thermal conversion of thermally thick biomass particles. Fuel Process Technol 95:96–108

Mehrabian R, Shiehnejadhesar A, Scharler R, Obernberger I (2014) Multi-physics modelling of packed bed biomass combustion. Fuel 122:164–178

Ndiema CKW, Mpendazoe FM, Williams A (1998) Emission of pollutants from a biomass stove. Energy Convers Manag 39:1357–1367

Saastamoinen JJ, Taipale R, Horttanainen M, Sarkomaa P (2000) Propagation of the ignition front in beds of wood particles. Combust Flame 123:214–226

Sami M, Annamalai K, Wooldridge M (2001) Co-firing of coal and biomass fuel blends. Prog Energy Combust Sci 27:171–214

van Loo S, Koppejan J (eds) (2008) The handbook of biomass combustion and co-firing, Earthscan, Washington, DC

Wendt JF (ed) (2009) Computational fluid dynamics: an introduction, 3rd edn. Springer, Berlin

Williams A, Pourkashanian M, Jones JM (2001) Combustion of pulverised coal and biomass. Prog Energy Combust Sci 27:587–610

Yang YB, Ryu C, Khor A, Yates NE, Sharifi VN, Swithenbank J (2005) Effect of fuel properties on biomass combustion Part II. Modelling approach-identification of the controlling factors. Fuel 84:2116–2130

Yin C, Rosendahl LA, Kaer SK (2008) Grate-firing of biomass for heat and power production. Prog Energy Combust Sci 34:725–754

Yin C, Kaer SK, Rosendahl L, Hvid SL (2010) Co-firing straw with coal in a swirl-stabilized dual-feed burner: modelling and experimental validation. Bioresour Technol 101:4169–4178

Zhou H, Jensen AD, Glarborg P, Jensen PA, Kavaliauskas A (2005) Numerical modeling of straw combustion in a fixed bed. Fuel 84:389–403

Chapter 7
Biomass Combustion: Carbon Capture and Storage

Abstract This chapter deals with the capture and storage of carbon dioxide produced by the combustion of biomass. Since biomass combustion is potentially carbon neutral, this technique could provide a method of reducing the carbon dioxide concentration in the atmosphere. The general features of this future technology of 'bioenergy carbon capture and storage', known as BECCS, are outlined. These cover biomass combustion with air or oxygen, gasification and chemical looping. Carbon capture and storage processes are briefly discussed.

Keywords Oxyfuel combustion · Carbon capture · Carbon storage

7.1 Introduction

The combustion of carbonaceous fuels such as biomass and fossil fuels results in the formation of carbon dioxide as one of the major products. The increase in carbon dioxide concentrations in the atmosphere has been linked with climate change and so considerable efforts have been devoted to reducing the emissions. This can be by increases in efficiency and abatement techniques although they have limited success up to the present time (IPCC 2013). As shown in Fig. 1.1, the combustion of fossil fuels and biomass will continue to be the major energy source for many decades and so considerable efforts have been made to develop techniques to capture carbon dioxide from the combustion exhaust products. This would thus nullify the effects of the combustion-generated carbon dioxide.

Research has indicated that it is desirable to keep global temperature increases below 2 °C compared with 1990 levels in order to avoid major climate change issues. However there are indications that this may not to be possible. Therefore large scale solutions may be necessary now such as geoengineering (Royal Society 2009). One such carbon-negative solution is the combustion of biomass combined with carbon capture and storage. Various terms have been used to describe this technology, initially Bio-CCS, but currently Bioenergy with Carbon Capture and Storage (BECCS) is the one most widely used.

© The Author(s) 2014

J.M. Jones et al., *Pollutants Generated by the Combustion of Solid Biomass Fuels*,
SpringerBriefs in Applied Sciences and Technology,
DOI 10.1007/978-1-4471-6437-1_7

A recent study stated that, 'globally, BECCS could remove 2 Gt of CO_2 from the atmosphere every year by 2050 using available sustainable biomass and possibly up to the equivalent to a third of all current global energy-related emissions. In Europe, BECCS could remove 800 Mt of CO_2 from the atmosphere every year by 2050 using available sustainable biomass—equivalent to over 50 % of current emissions from the EU power sector' (ZEP 2012). The development of biomass combustion together with carbon capture and storage would have the effect of reducing the concentration of carbon dioxide in the atmosphere, and potentially could have a beneficial effect on the climate. The IPCC have recently concluded that BECCS could play an important role in many low-stabilization scenarios where it is typically used in response to overshoots in levels of CO_2 accumulation in the atmosphere (IPCC 2014). However this could only be achieved in large scale plant and the power generation industry or other large industrial complexes. Since large scale chemical engineering operations are involved it might be argued that the most appropriate option is a power generation complex coupled with an associated chemical plant. This would be greatly enhanced if the carbon dioxide product could be utilised as a chemical feedstock. Clearly this would be a long-term objective.

As with coal-CCS there are several ways forward (Edge et al. 2011, 2012; Gibbins and Chalmers 2008; Toftegaard 2010). One is by oxy-biomass combustion and the other is by gasification. In both cases the carbon dioxide is removed from the reacted gases.

7.2 Combustion with Air or Oxygen

When biomass is burned with air the CO_2 produced can be removed from the flue gas, this is commonly termed 'post-combustion capture'. Many industrial scale technologies exist for CO_2 capture including absorption, adsorption, membranes and cryogenic technologies. Many of these technologies are still emerging at pilot scale; however absorption using an amine solvent is commercially developed and is undergoing large scale trials. This method can be used when CO_2 concentrations in flue gas are low.

Oxyfuel combustion however is a strong contender to be the preferred technology and this can be undertaken using pulverised biomass or a fluidised bed. Most of the R&D has been directed to the former case and only that is being considered here except for the special case of loping that is described later. The biomass fuel is burnt with oxygen and recycled flue gas (RFG) instead of air. RFG reduces the combustion temperatures and ensures a flue gas with high CO_2 concentrations of up to 95 % is produced. The recycle arrangements and definition are shown in Fig. 7.1.

The recycle rate of the combustion gases can be defined in a number of ways which are based on the flow of input oxygen and the amount of flue gas returned to the boiler.

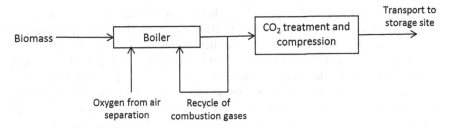

Fig. 7.1 Outline of oxy-combustion plant

A number of pilot scale electrical generation plants using coal are coming into operation at present and since co-firing with biomass is a mature technology this could be readily be adapted to at least part bio-CCS. Large-scale utility boilers are considered the highest potential targets for CCS with fossil fuels in the early CCS commercialisation phase. Co-firing of biomass could realise negative emissions in many of these utilities and is reasonably easy to implement.

The CO_2 captured can be transported and stored following treatment to remove water and acid gases. Overall the major challenge is the efficiency penalty associated with the production of the oxygen, the CO_2 capture stage and the compression of the CO_2. There are issues in using oxygen in place of air in biomass combustion. The first is the radiative properties of the combustion gases because the main component N_2 has been replaced by the infra-red active carbon dioxide and water vapour (Black et al. 2013). Because of this change the specific heat of the gases change as will the resultant flame temperature unless the fuel/air ratio is changed but usually the recycle rate is adjusted to match the heat transfer required. Usually R increases with the gas emissivity, which is the size of the furnace. Usually the oxygen concentration is increased to 25–30 mol%. Biomass flames commonly have reduced temperatures, as the plant is derated to prevent deposition from biomass ash components. Another factor is the fact that the devolatilisation of biomass particles occurs to give about 90 % gaseous products with little char or soot. Thus the radiation from particulate in biomass flames is reduced compared with coal flames.

There are also other factors that come into play. The residence time changes as does the average HHV of the gases. There is no practical experience of plants of this nature although some theoretical studies have looked at the required changes to the recycle ratio to maintain boiler heat transfer and boiler exit temperatures.

Very many studies have been made of the physics and chemical kinetics of biomass combustion in O_2/CO_2 atmospheres using laboratory methods such as TGA, drop tube reactors and flame reactors, e.g. Álvarez et al. (2013), Hecht et al. (2013). The replacement of nitrogen by carbon dioxide has a significant effect on the specific heats and radiative and transport properties of the reacting gas, Whilst it is not expected that the kinetics of the initial pyrolysis step will be changed significantly by the different gas composition (other than temperature changes), there will be significant changes in the reaction of the volatiles released and of the char combustion (Gil et al. 2012; Toftegaard et al. 2010). In the case of coal combustion where there is a significant amount of char produced the additional oxidation reaction by the presence of carbon dioxide is about 10 % (Hecht et al. 2013). But biomass devolatilisation produces much less char and the effect will be accordingly smaller.

There is also another variant of oxy-fuel combustion possible. In this steam is used to moderate the flame temperature (Seepana and Jayanti 2010). In this process oxygen is mixed with steam and the mixture is used for combustion with biomass fuel. The advantage of this method is that flue gas recirculation is avoided and the volumetric flow rates through the boiler and auxiliary components is reduced by about 39 % when compared to the conventional air-fired coal combustion power plant leading to a reduction in the size of the boiler. The flue gas, after condensation of steam, consists primarily of CO_2 and can be sent directly for compression and sequestration. This may be advantageous with biomass because of the high oxygen content in the fuel.

7.3 Gasification

Gasification of biomass can be used and in this application is frequently termed 'precombustion capture' Biomass is reacted usually with oxygen to produce a synthesis gas of carbon monoxide (CO), hydrogen (H_2), CO_2 and H_2O. A water shift reactor is used to produce CO_2 and H_2. H_2 can be used in a gas turbine combined cycle plant and CO_2 can be captured. The term 'Integrated gasification combined cycle (IGCC)' has been used for this cycles when coal is used as the initial fuel (Basu 2013).

The difficulty with using biomass with this process arises from the low CV of the feedstock, and from the tar and inorganic deposits that can be produced if fluidised beds are used or the extreme corrosion that can occur in higher temperature reactors.

7.4 Chemical Looping

Chemical looping combustion uses two linked fluidized bed reactors where a metal oxide is employed as a bed material providing the oxygen for combustion in the fuel reactor. The reduced metal is then transferred to the second bed (air reactor) and re-oxidized before being reintroduced back to the fuel reactor completing the loop. This uses oxygen and so the flue gas consists of carbon dioxide and water vapour as well as trace pollutants.

As an example, a nickel based system burning pure carbon would involve the two redox reactions:

$$2Ni(s) + O_2(g) = 2NiO(s) \qquad (R\ 7.1)$$

$$C + 2NiO(s) = CO_2 + 2Ni(s) \qquad (R\ 7.2)$$

The overall reaction is simply oxidation of the carbon with the nickel acting as a catalyst. The difficulty with using biomass as a feedstock for this process is the low carbon content in the feedstock necessitating the use of very large reactors. The high biomass oxygen content may be an issue too.

7.5 Carbon Capture and Storage

Once the carbon dioxide is separated from the flue gas it has to be purified to remove water and acid gases and transported and stored (Rackley 2009). For large scale transportation the only option appears to be by pipeline which is a well-established technology. As far as CO_2 storage is concerned the options are (ZEP 2012):

- Deep saline aquifers (saltwater-bearing rocks unsuitable for human consumption)
- Depleted oil and gas fields (with the potential for Enhanced Oil/Gas Recovery)
- Deep un-mineable coal beds (with the potential to extract methane)

There are a number of difficult issues.

1. The high cost of storage at a very large scale and the way in which this can be financed so as to make it viable.
2. The energy penalty involved from parasitic loading of capture and regeneration equipment.
3. The uncertainty of reliable geological storage for CO_2 (Gluyas and Mathias 2013).
4. Long-term ownership of stored carbon dioxide.

These problems can be simplified into two factors, which require to be solved in a way which is credible by industry and public, for timescales beyond the lifetime of individual Overall is the prospect of Bio-CCS is a major challenge but with considerable future benefits.

References

Álvarez L, Gharebaghi M, Jones JM, Pourkashanian M, Williams A, Riaza J, Pevida C, Pis JJ, Rubiera F (2013) CFD modeling of oxy-coal combustion: prediction of burnout, volatile and NO precursors release. Appl Energy 104:653–665

Basu P (2013) Biomass gasification, Pyrolysis and Torrefaction practical design and theory. Academic Press, Waltham

Black S, Szuhánszki J, Pranzitelli A, Ma L, Stanger PJ, Ingham DB, Pourkashanian M (2013) Effects of firing coal and biomass under oxy-fuel conditions in a power plant boiler using CFD modeling. Fuel 113:780–786

CRECK (2014) http://creckmodeling.chem.polimi.it/

Edge PJ, Gharebaghi M, Irons R, Porter R, Porter RTJ, Pourkashanian M, Smith D, Stephenson PL, Williams A (2011) Combustion modeling opportunities and challenges for oxy-coal carbon capture technology. Chem Eng Res Des (Official J Eur Fed Chem Eng A) 89(9):1470–1493

Edge PJ, Heggs PJ, Pourkashanian M, Stephenson PL, Williams A (2012) A reduced order full plant model for oxyfuel combustion. Fuel 101:234–243

Gibbins J, Chalmers H (2008) Carbon capture and storage. Energy Policy 36:4317–4322

Gil MV, Riaza J, Álvarez L, Pevida C, Pis JJ, Rubiera F (2012) Kinetic models for the oxy-fuel combustion of coal and coal/biomass blend chars obtained in N_2 and CO_2 atmospheres. Energy 48:510–518

Gluyas J, Mathias S (2013) Geological storage of carbon dioxide. Woodhead Publishing, Cambridge

Hecht ES, Shaddix CR, Lighty JS (2013) Analysis of the errors associated with typical pulverized coal char combustion modeling assumptions for oxy-fuel combustion. Combust Flame 160:1499–1509

IPCC (2013) Climate change 2013: The physical science basis. In: Stocker TF et al (eds) Contribution of working group I to the fifth assessment report of the intergovernmental panel on climate change. Cambridge University Press, Cambridge

IPCC (2014) Climate change 2014: mitigation of climate change. www.ipcc.ch

Rackley S (2009) Carbon capture and storage. Butterworth, Heinemann. ISBN 9781856176361

Royal Society (2009) Geoengineering the climate science, governance and uncertainty. Royal Society, Sept 2009

Seepana S, Jayanti S (2010) Steam-moderated oxy-fuel combustion. Energy Convers Manag 51:1981–1988

Toftegaard MB, Brix J, Jensen PA, Glarborg P, Jensen AD (2010) Oxy-fuel combustion of solid fuels. Prog Energy Combust Sci 36:581–625

ZEP (2012) EU zero emissions platform—biomass with CO_2 capture and storage (Bio-CCS)—the way forward for Europe, 2012. www.zeroemissionsplatform.eu

Appendix A: Calculation of Flue Gas Composition

Consider 100 kg of biomass (pine wood) with an Ultimate Analysis on a dry ash free basis given in Table A.1.

This calculation assumes that all the carbon is burned i.e. there is no carbon in the ash or carbon monoxide formed. If there is the unburned carbon has to be subtracted from the value of carbon in the above table. Also that the S, N and Cl are negligible and there is no ingress of N_2 in the flue or sampling system. Moisture content given in the Proximate Analysis has to be allowed for in the calculation as well.

Assume the combustion reactions are

$$C + O_2 = CO_2$$

$$2 H_2 + O_2 = 2 H_2O$$

Oxygen required for stoichiometric combustion = $4.33 + 1.6 - 1.3 = 4.63$ kg-mol

$$\text{Air requirement} = [4.63 \times 100 \times 22.41]/[21] = 494.1 \text{ m}^3 \text{ at NTP}$$

Composition of the **dry flue gas** from the stoichiometric combustion of 1 kg of biomass:

N_2 from theoretical air = $4.94 \times 0.79 = 3.90$ m^3.
N_2 from biomass = $0.001 \times 0.224 = 0.0002$ m^3.
CO_2 from biomass = $4.33 \times 0.224 = 0.97$ m^3.
Total amount of **dry flue gas** per 1 kg wood = 4.872 m^3.

If the concentration in the wet flue gas is required then the water content has to be included in the above calculation.

© The Author(s) 2014
J.M. Jones et al., *Pollutants Generated by the Combustion of Solid Biomass Fuels*,
SpringerBriefs in Applied Sciences and Technology,
DOI 10.1007/978-1-4471-6437-1

Table A.1 Calculation of flue gas composition

Element	Molecular weight	Mols O_2 required
C, 52.00	12	4.33
H, 6.20	2	1.6
O, 41.60	32	−1.3
N, 0.2	28	

Therefore % CO_2 = 0.97/4.872 = 19.91 assuming stoichiometric combustion (0 % O_2).

For 20 % excess air = 16.6 % CO_2
For 40 % excess air = 14.2 % CO_2
For 50 % excess air = 13.27.

Appendix B: Gaseous Emissions Conversion Table

A Conversion Chart is given below. The unit used is the Normal m^3 (Nm^3) which is commonly used by metric countries and the power industry generally. The reference temperature is 0 °C and 1 bar pressure. Other countries, such as the US EPA, use 20 °C and some tables and converters available on the web use 25 °C. Corrections have to be applied at altitudes above sea level. Note that NO$_x$ is treated as NO$_2$.

To convert from 1 ppmv	To mg Nm^3
NO$_x$	2.05
NO	1.34
CO	1.25
SO$_2$	2.86
HCl	1.63
NH$_3$	0.76
HCN	2.86

© The Author(s) 2014
J.M. Jones et al., *Pollutants Generated by the Combustion of Solid Biomass Fuels*,
SpringerBriefs in Applied Sciences and Technology,
DOI 10.1007/978-1-4471-6437-1

Appendix C: Physical and Thermal Properties of Biomass

Some typical values are given here, but they depend on the moisture content and the exact structure of the biomass.

Pine	
Density	550 kg/m^3
Thermal conductivity	0.12 W/mK
Heat capacity	1670 J/kgK
Straw	
Density	400 kg/m^3
Thermal conductivity	0.1 W/mK
Heat capacity	350 + 2T J/kgK

© The Author(s) 2014
J.M. Jones et al., *Pollutants Generated by the Combustion of Solid Biomass Fuels*,
SpringerBriefs in Applied Sciences and Technology,
DOI 10.1007/978-1-4471-6437-1